# THE
# HEALING
# CODE OF
# NATURE

## Also by Clemens G. Arvay

*The Biophilia Effect: A Scientific and Spiritual Exploration of the Healing Bond Between Humans and Nature*

# THE
# HEALING
# CODE OF
# NATURE

## Discovering the
## New Science of
## Eco-Psychosomatics

## CLEMENS G. ARVAY

Translated by Victoria Goodrich Graham

*Der Heilungscode der Natur*
*Die verborgenen Kräfte von Pflanzen und Tieren entdecken*

SOUNDS TRUE
BOULDER, COLORADO

This book is not intended as a substitute for the medical recommendations of physicians, mental health professionals, or other health-care providers. Rather, it is intended to offer information to help the reader cooperate with physicians, mental health professionals, and health providers in a mutual quest for optimum well-being. We advise readers to carefully review and understand the ideas presented and to seek the advice of a qualified professional before attempting to use them.

Published 2018

Cover design by Lisa Kerans
Book design by Beth Skelley

Printed in Canada

Library of Congress Cataloging-in-Publication Data

Names: Arvay, Clemens G., author.
Title: The healing code of nature : discovering the new science of eco-psychosomatics /
    Clemens G. Arvay ; translated by Victoria Goodrich Graham.
Other titles: Heilungscode der Natur. English
Description: Boulder, Colorado : Sounds True, [2018] | "Originally published as Der Heilungscode der Natur. Die verborgenen Kräfte von Pflanzen und Tieren entdecken. 2016, Riemann Verlag, a division of Verlagsgruppe Random House GmbH, München, Germany."—Title page verso. | Includes bibliographical references and index.
Identifiers: LCCN 2017050617 (print) | LCCN 2017050982 (ebook) | ISBN 9781683640325 (ebook) | ISBN 9781683640318 (paperback)
Subjects: LCSH: Naturopathy. | Nature, Healing power of.
Classification: LCC RZ440 (ebook) | LCC RZ440 .A78713 2018 (print) | DDC 615.5/35—dc23
    LC record available at https://lccn.loc.gov/2017050617

10 9 8 7 6 5 4 3 2 1

"All life processes, from a cell all the way to the coexistence of humanity and nature, are always profoundly intertwined. All parts mesh with each other."

**HERMANN HAKEN**[1]

# CONTENTS

# NATURE AND HEALTH – A TOPIC FOR THE NEW MILLENNIUM

Books about healing are often written with older audiences in mind. There are, however, good reasons to also address younger people, who will be able to approach open questions from a new angle and help shape the future of our society. In the 1980s and 1990s, the so-called Millennials were born, also known as Generation Y (as in "why"). The Millennials (but not just them) really do question almost everything in existence: Does our lifestyle have to harm the environment? Is there really nothing more than what science has discovered so far?

Henry David Thoreau (1817–1862) answered these questions through the construct of *simplicity*: an independent life close to nature (as an alternative to a society controlled by artificial needs) brings humans closer to real life.

The Millennials are sensitized to the protection of natural habitats and know the deficits of a purely technological, rationalistic view of nature. Therefore, they are particularly open to the increasingly evidence-based body of knowledge about the positive effect of nature on humans. (Clemens G. Arvay, born in 1980, belongs to the oldest age group of the Millennials.) But middle-aged and elderly people are also familiar with the problem: the world is economically drained, and the destruction of its natural resources is just as alarming, if not more so.

Arvay shows how closely the health of the individual and society is related to the health of the planet. The main focus of this book is comprehensively explaining *how* contact with nature affects our organs and cells. The author also highlights our relationship to animals and shows how encounters between humans and animals have proven medical and therapeutic effects on us.

In this book, Arvay expands on the basic insights from his book *The Biophilia Effect*, which focuses on showing us how to experience the healing powers of the forest through practical exercises. He provides abundant new evidence for the wide range of nature's positive effects and elaborates on the healing powers of plants and animals with a solid scientific basis. As a biologist, he makes it clear that "green science" plays an important role in the development of preventive measures and treatments in medicine. We have implemented these findings at our university in order to improve patients' lives in the neighboring geriatric center through garden therapy, for example. The successful results show that this way of experiencing nature's resources can be beneficial for everyone.

I hope that you will gain many new and useful insights while reading this interesting book and find inspiration to discover how you can help maintain or restore your health responsibly through the healing power of plants and animals.

**Dr. Thomas Haase**
Rector of the University College for Agrarian
and Environmental Pedagogy
*Vienna, February 2016*

# WHAT AWAITS YOU IN THIS BOOK

Over the course of this book, I will "carry you away" on a journey back to the first living cell on Earth to explore our profound connection with plants and animals and to develop a medicine of the future from these insights. This medicine will see us humans, once again, as we are: as natural beings, inseparable from our natural habitats. We do not end at the surface of our skin!

I call this science "nature-human medicine," but in the course of the book, I will introduce the more suitable term "eco-psychosomatics." I will come to this term step by step because it needs to be built upon a solid scientific base. We'll begin our exploration with some astonishing discoveries of modern research. Did you know that the sight of a tree alone activates the self-healing powers of a human being and that more trees in big cities would result in measurable, rejuvenating changes in the city-dwellers' blood, making people look and feel years younger? Contact with friendly animals is proven to strengthen our immune system and even helps with the recovery of critically ill patients. At some clinics, dogs, cats, guinea pigs, and rabbits are already employed as "therapists"—with wonderful results! We'll also see how encounters with wild animals can have a healing effect.

Did you know that when we inhale a cocktail of bioactive plant matter from the forest air, it strengthens our bodies' defenses so much that even the international cancer research community has started paying attention to these tree substances? You will learn what these substances have to do with the "language" of plants.

Nature-human medicine is anything but esoteric. It is the key to better medical treatments because it takes into consideration that people have been connected with plants, animals, and ecosystems over eons. By considering this, we can better understand diseases, protect ourselves from them, and treat them more effectively. We also become more familiar with our own species and our relationship with animals, which are more similar to us than many of us have realized. Modern discoveries in biology show that mechanisms are in place in our brain and nervous system that support us in establishing true friendships and relationships with animals. These biochemical connections trigger beneficial effects on our health—and, coincidentally, on the animals' health as well, as long as we treat them with dignity. The upcoming chapters also provide arguments to rethink society's treatment of animals.

Everything in this book revolves around scientific explanations for the hidden powers of plants and animals, which keep us healthy when we open up to them. We will be exploring the healing effects of nature down to our organs and deep into our cells. Thus, we will see, for example, how our nerves translate the symbolic language of nature and sensory stimuli from the world of plants and animals into something physical.

With that in mind, I welcome you now to join me in exploring nature-human medicine. Our journey will take us all the way back to the Big Bang.

# 1

# THE MYSTERY OF TREES

## What Do Trees Have to Do with Our Health?

When I was writing this book, my little son, Jonas, was in the hospital, and I was at his side as his caregiver. We shared a room at the university clinic for pediatrics in my former hometown of Graz in the south of Austria. The inpatient treatment lasted several weeks and was associated with numerous hardships for my little patient, who was then only seventeen months old. At times they pricked his finger every day to draw blood, and during the night, he was often attached to an IV for hours or connected with wires to medical devices. The long hospital stay and the constant uncertainty of how long it was going to last drained my strength, as well as that of my son. But around the clinic there was something that brightened our mood and helped us to not lose courage: a vast forest. The pediatric clinic, which lay directly on the edge of the forest, was literally embedded in a woodland landscape.

Every day, as soon as we went outside through the ground-floor entrance, Jonas started to kick his legs and laugh in his baby carriage, full of joyful anticipation. We had discovered a narrow path off the main route in the forest at the beginning of our stay. Right around the first bend, we were greeted by an old beech tree with a thick, silver-gray trunk. Every time we passed it, Jonas followed the trunk up to the top with his gaze. Above our heads, the beech spread its branches into a mighty crown. It was autumn, and the foliage shone in intense colors of red and yellow. We roamed through the thicket and picked the last berries of the year from twisting, thorny plants. We watched deer and squirrels. Once we even saw a fox.

I was impressed how every day something new in this forest fascinated my little boy, so even after his exhausting experiences in the hospital, he could still laugh out loud with me in nature. I am sure the regular breaks in the woods helped my son gain some distance from the daily hospital routine and allowed him to process stressful experiences more easily, thereby saving his psyche from deeper scarring.

I felt how the forest had the same effect on me. I returned from the woodland with renewed strength every time. We definitely would have been worse off in a hospital surrounded by concrete and without any green space. Other patients I met in the woods talked about the same beneficial effects they experienced there. The nurses, doctors, and psychologists also used the forest during their breaks or after work to take a step back from their often stressful jobs. One oncologist, who daily treats children with cancer, told me that her evening walks in the woods helped her to deal with her own feelings of sadness, which are as much a part of her profession as happiness after successful treatments.

Even for patients who cannot leave the hospital, the environment is extremely important. Environmental scientist Rachel Kaplan at the University of Michigan found in a scientific investigation that even glancing briefly through the window at trees or green areas relieves and relaxes us in stress situations. I experienced this phenomenon in the clinic as well. From my son's room, we could see through the window over the crown of the forest. Again and again I caught a soothing glimpse of it, which gave me the opportunity to "escape" our turbulent life in the hospital and take a deep breath. This also made me look forward to our next opportunity to visit the forest; its broad, leafy canopy symbolized the "world out there" and the prospect of leaving the daily hospital routine behind me. A cement wall would not have had the same effect.

Scientists have been comparing the effects of trees versus brick walls near hospitals for more than thirty years. In April 1984, *Science*, one of the world's best-known scientific journals, published the results of a clinical trial that researchers conducted on a variety of patients over a number of years. Health-care design researcher Roger Ulrich, who teaches and researches as a professor at Swedish as well as US universities, demonstrated together with his colleagues that the mere view from a hospital window influences healing. The patients in the study all required routine gallbladder surgery. The treatment and accommodations were identical for all of them, with only one factor modified: Some of the patients could look out of the window at a tree, which Professor Ulrich referred to as the "tree group." The other patients looked through the hospital window at nothing but a brick wall. The results speak for themselves: The patients from the tree group were able to return home more quickly than those from the brick-wall group since wound healing and overall recovery were accelerated. They also needed significantly fewer, and lower-strength, painkillers.[1] Finally, there were even fewer postoperative complications in the tree group, although Ulrich explains that he considered this effect to be secondary and attributed it to the weaker pain medication causing fewer side effects.

While I was with my son in the hospital, I kept thinking about this study and was happy that we could not only see one tree but a whole forest when we looked out his window.

The way in which Professor Ulrich pursued his investigations corresponds to a research method with comparative groups that is frequently used in clinical trials. The scientists divide their patients into groups. They only change a single factor in each group, in this case "tree" or "brick wall." The effects of these variables were being tested. Ulrich's participants had no idea that the tree or brick wall in front of their windows was part of an experiment; in other words, they were "blind" to these variables. Then the researchers compared the two groups on average. What changes does the tree cause? What changes with the wall? In Ulrich's experiment, the physicians did not know that the trees and walls outside the windows had any meaning either.

They, too, were "blind" to it. This is called a "double-blind study." In this way, researchers could rule out that the mere expectations of the patients had an unconscious influence on their healing or that the doctors accidentally behaved differently to the tree patients than the wall patients because of their own expectations.

These double-blind studies are the gold standard in medical research. They are also used in testing medicines. Each drug must be compared to a placebo with no active ingredient, and neither the doctors nor the patients usually know which pill has the active substance. As Ulrich's study showed, the influence of plants on people's health can also be determined by using such comparative studies. Group comparisons will continue to play an important role in this book as we take a closer look at the scientific background of nature-human medicine.

Since Ulrich's tree studies, research has provided much more evidence that trees have a surprisingly positive effect on our health that goes far beyond the mere "feel-good effect." Qing Li, a professor of medicine from the Nippon Medical School in Tokyo, and his research team analyzed statistical health data of the Japanese population and concluded that significantly fewer people died of cancer in forested areas than in areas with no forest. In this study, the scientists included all prefectures in Japan.[2]

This process is entirely different from clinical double-blind studies. Qing Li and his colleagues conducted an epidemiological study, which is the statistical study of the health status of large populations where the scientists do not directly examine their "patients" but instead access databases. Thus, they can capture and compare information from very large parts of the population. Through complex statistical procedures, Li's team had to ensure that the variable "forest area" was at the center of their study. Epidemiologists use anonymous data from health authorities and health insurance funds. How often does a person visit doctors or have to go to the hospital? What are their diseases? What drugs do they need to take? Qing Li and his team were able to link the incidence of cancer to proximity to vegetation. Because they studied the entire country, it is clear that trees—both in the city and in the country—reduce the risk of dying of cancer. Epidemiological studies are widely used in health sciences and are scientifically recognized. They are an excellent way of checking the healing powers of nature on large populations.

Marc Berman, an assistant professor of psychology at the University of Chicago with a focus on environmental neuroscience, went one step further. He not only evaluated health data, but also combined it with satellite images and tree mapping from Canada's largest metropolis, Toronto. This extensive epidemiological study involved a large team, and its results were reported online in 2015 by the renowned scientific journal *Nature*, published by Springer.[3] It turned out that the city inhabitants' state of health improved as the number of trees around their neighborhoods increased. Berman showed that the more trees that grow near people living in urban areas, the lower the statistical risk those people have of suffering from heart disease, diabetes, hypertension, and other typical "lifestyle diseases." This study did not include parks and green spaces; the health effects were only from those trees that were part of the urban streets of Toronto, such as trees along avenues, shrubs along the roadside, and small traffic islands with trees in the big city traffic. Berman and his team recorded half a million urban trees. They also compared the medical effects of these trees with other influences on our health. This is when the trees' potential really became evident. According to Berman and his coworkers, ten additional trees around a block of homes would have the same effect on the inhabitants' health as being seven years younger. This means the healing effects of nature are by no means confined to forests, but can also be found in large cities.

In an interview with the *Toronto Star*, Faisal Moola, professor of forest ecology at the University of Toronto and a coauthor of the study, was very pleased that, once again, environmental factors were proved to play an important role in our health. This reality has been underestimated by political decision makers so far. He demanded more trees in Canada's major cities. Glenn De Baeremaeker, city councillor in Toronto, considered the tree study groundbreaking because "it's a pretty magical solution, for peanuts."[4]

## Trees and Our Immune System: "We Need to Understand"

Berman was also enthusiastic about the results and speculated on how trees affect our health. "Is it that the trees are cleaning the air? Is it that they are encouraging people to go outside and exercise more? Or is it their aesthetic beauty? We need to understand that."[5]

Berman was certainly not wrong when he thought about the air-purifying effect of the trees. Additionally, green elements in the city are surely a motivation for more outdoor activity, and almost everyone would agree on the aesthetic value of trees. Yet that is not enough to explain why more trees in the city make us feel years younger and protect us from common diseases. In order to *understand* the healing relationship between humans and trees, we first need to learn more about what the doctors from Tokyo have found out. These findings open up completely new approaches and directions of thought in our search for explanations for the healing effects of trees.

In Japan, public funds are available for research in nature-human medicine. The Japanese government also finances elaborate studies with expensive testing methods that run for many years. Doctors integrate contact with nature into therapies offered at public hospitals, and experts study the health aspects of trees and forests at medical universities.

The tradition of *shinrin-yoku* ("forest bathing") goes back a long time in Japan's folk medicine.[6] North and South Korea also have the same tradition, but it's called *sanrimyok* there. But forest bathing has its longest history in China and Taiwan and has been called *senlinyu* there for centuries. The largest official senlinyu resort is situated in the north of Taipei, the capital of Taiwan. Ancient knowledge about healing from nature is also found in Traditional Chinese Medicine. Numerous exercises from qigong, a meditative movement system that also contains elements of martial arts, are designed to "absorb the chi of nature" as Traditional Chinese Medicine puts it. These exercises are carried out mainly in forests or green areas with trees. Even the qigong masters of the past apparently knew that nature not only heals in the form of plant- and mineral-based pharmaceutical substances, but also by a person simply being present in a green space and breathing. In qigong, absorbing the chi of nature is always associated with intense breathing techniques. Furthermore, naturopathic physicians in Japan not only translate the term shinrin-yoku as "forest bathing," but also often as "inhaling the forest atmosphere." This suggests that something must be in the forest air that medically affects us when inhaled.

In 2013 Qing Li published the results of several studies that he had conducted with Japanese, Korean, and Chinese scientists in previous years. They discovered that spending time in forests leads to changes in the human body, which are clearly detectable via blood tests and other medical evaluations. At this point, I am going to focus on only one example of these changes, namely, the effect of trees on the human immune system.

Qing Li and Tomoyuki Kawada, a physician in environmental medicine, tested their subjects' blood before and after visits to a forest. They found that the forest activates and strengthens important components of the human immune system. Our immune system responds to even short walks in the forest. A whole day in the woods resulted in an almost 40 percent increase in natural killer cells in the subjects' blood on average. Not only were there more of these important defense cells, but they were also much more active than before. The effect lasted for seven days. After spending two days in a forest, the participants' blood contained 50 percent more killer cells than before, and the effect was still measurable thirty days later.[7]

The term "natural killer cells" does not exactly evoke the most pleasant associations. The warlike metaphors with which scientists describe the processes in our immune system sometimes trigger debates about whether such a choice of words is justified. Can biological processes be compared with war? There is often talk of "weapons" that our body uses against "invaders" (germs) and "enemies" (tumors). Some scientists fear that it might be a strain on patients to imagine their body as a battlefield. Others, on the other hand, believe that it encourages a fighting spirit when we get sick. Statistically speaking, patients with a fighting spirit mindset are more likely to survive serious illnesses than patients who feel powerless against the disease.[8]

Either way, the choice of words is not too far-fetched. The natural killer cells, which the human body produces more of in the forest, are responsible for rendering viruses in our body harmless. They keep the viruses in check until enough special antibodies have been formed that can kill all of them. In addition, the natural killer cells attack dangerous cells in our body that could mutate into malignant tumors. They therefore keep us healthy, protect us preventively against cancer, and are indispensable for healing in this disease.

A look at the way natural killer cells work against viruses and tumor cells makes the combative word selection from the textbooks of immunology even more apt. Natural killer cells are part of our innate immune system. That is, they are not special antibodies that only appear—and then attack—a particular virus after contact, for example a flu virus. The natural killer cells form in our bone marrow and are already active when we are born. They accompany us all our lives. A killer cell has the ability to recognize when a body cell is infected by a virus based on its surface structure. It then activates a process that immunologists call the "killing machine."[9] Since viruses cannot be directly attacked, the entire body cell has to be rendered harmless. Viruses inject their own genetic material into that of the cell, thereby causing the cell to reproduce new viruses "by default." The killer cells deposit microscopically small granules on the surface of the infected cell. These granules contain proteins that penetrate the cell and initiate natural cell death. In other words, they force the cell to self-destruct, which also means the end of the virus.

Natural killer cells do the exact same thing to the very dangerous mutating cells that lead to cancer. Cancer always begins with one cell that should have died but didn't. It generates more immortal cells that begin to proliferate. Killer cells help if a cell does not want to die and constitutes a health hazard for humans. The granules and proteins required by the killer cells to do this are also called "anticancer proteins" because of their function.[10]

Qing Li and Tomoyuki Kawada found out that "Doctor Forest" not only gives rise to additional and more active killer cells, but also promotes the formation of these anticancer proteins. Subjects of the study had significantly more of them in their blood after spending time in a forest. Simply being present in the forest can therefore support our natural mechanisms that are responsible for the defense against viruses and protect us from cancer or fight against existing tumors. It is clear that these effects are directly attributable to the forests because the researchers didn't note any positive changes in comparison groups that were in urban areas. The participants did not exercise in the forest, which could also have triggered the health-promoting effect; they were simply present in the forest.

Such studies are all about group comparisons, which I have already described as the gold standard of medical research. One group goes into the woods; the other goes into the city. There is only one catch when scientists send their patients directly into nature: in the case of a tree in front of a hospital window, researchers can conceal that it plays a central role in the experiment. For field studies in nature, however, it is very difficult to keep participants in the dark about what is being compared. A double-blind study, as I described it, is often not possible in this type of field study. However, researchers can create comparable conditions between the "city group" and the "nature group" and meet scientific requirements. All subjects must spend the same amount of time in nature or the city and be active in a comparable way. Most of the time, the scientists leading the study even ensure that all of them eat the same. Only the environment is different between the two groups.

By taking blood samples in nature, it is possible to combine field studies with laboratory tests. Natural killer cells and anticancer proteins of the immune system are often used by doctors as conclusive biomarkers. Therefore, we are also in line with scientific trends when we use these important parameters of our immune system in order to prove the beneficial effects of nature.

An accompanying study revealed an additional positive effect that trees have on the human immune system. Our body has "first-aid cells" that are designed to attack invading pathogens as quickly as possible while our body mobilizes other defenses. These first responders are called "neutrophils" and swim in the bloodstream. With the help of sticky substances, they are able to instantly attach themselves to the inner walls of the blood vessels at any time and to pull themselves out of the bloodstream without being washed away. That way, they do not miss their target. They penetrate the body tissue and basically "eat" the intruders by absorbing and digesting them. We sometimes feel and see this process as an inflammation. After spending time in a forest, an increase can be detected in the sticky substances that first-aid cells use to pull themselves out of the bloodstream and eat pathogens.[11] This is further tangible evidence of the effectiveness of nature-human medicine.

In the case of ill health, spending time among trees does not constitute a substitute for adequate medical treatment but is rather a complementary measure. The term "complementary medicine" means that common medical treatments are *supplemented* by additional therapies, not used in competition with them. Nevertheless, when researching the healing effects of trees, we are dealing with a science that can draw on verifiable facts. Forest medicine is therefore "evidence-based," as it is called in modern medical jargon.

For now, let us continue looking at the effects of trees on the human immune system—at the natural killer cells, the first-aid cells, and the anticancer proteins. *What* exactly strengthens them in the forest? In order to answer this key question, several teams of biologists, physicians, and psychologists headed to the most beautiful forest areas in Japan, including the Nagano Prefecture, where the 1998 Winter Olympics took place.

## From Tree to Tree . . . and Our Cells Are Listening In

Equipped with measuring instruments, Tatsuro Ohira and Naoyuki Matsui from the Forestry and Forest Products Research Institute in the Japanese town of Tsukuba traversed the forests and documented exactly which substances they could detect in the air.

We all know how aromatic it smells under trees, yet the fragrance of woods, leaves, needles, and bark is much more than just odors in our nose. When we happily exclaim that "it smells so good in the forest," many of us are unaware that trees do not just smell good, but that the aromatic substances fulfill a very specific biological function. What floats into our nose as a fragrance is part of what plants use to communicate.

Biologists have known for decades that plants exchange information. In the media and in popular books, however, this knowledge is often sold as something new. They refer to "talking trees" and "murmuring ferns" that "whisper" to each other. According to one press report, there are apparently even "timid" plants. Of course, this anthropomorphic language doesn't work for plants, since they are quite different from us. They split off from animals in the tree of life a very long time ago, and they have gone a completely different way in their development. Plants are an entirely different life-form, and that is what makes them so fascinating. I have great respect for plants, but they do not whisper to each other and are not shy or forward either.

For the time being, let's keep in mind that trees, shrubs, other plants, and mushrooms exchange messages through gaseous substances. Microorganisms in the soil do the same. A forest is a place of lively communication with a multitude of different life-forms and a variety of interrelated organisms. This is true for every natural habitat. Aquatic plants, algae, corals (which belong to the animal kingdom), and other sea flora maintain their ecosystem by exchanging information and can even send messages to marine animals. Countless chemical compounds carrying meanings float through seawater. Likewise, they zip around in the forest air.

They all belong to a group of substances called *terpenes*. With their help, trees and other plants inform each other about pests that are attacking, for example. They can even communicate to one another details about the advancing pest, such as the nature of the pest and the strength of the attack.

Plants are able to lure useful insects via terpenes, which then act as opponents to pests. They also use terpenes to mark their territory and stop competing plants and hungry vermin by means of deterrent signals. Mushrooms use terpenes as attractants so that their reproductive cells find their way to each other. The fragrances of essential oils are also terpenes. We are about to see how studies have now demonstrated the healing power of plant-based oils. In nature, however, we do not even get close to perceiving all the terpenes with our sense of smell. Many remain hidden from our senses—but not from the measuring devices of scientists.

Japanese forest researchers quickly assumed that terpenes measured in high quantities in the forest could be responsible for the health-promoting effect of trees on humans.[12] To test the accuracy of their hypothesis, the researchers planned an elaborate experiment.

A team of eleven scientists from the Nippon Medical School in Tokyo invited numerous people to spend three nights in urban hotels. Using a nebulizer, researchers enriched the air in half of the participants' hotel rooms with various terpenes from the Japanese cypress. The oil of this tree contains many substances found in forest air, especially the terpenes called pinene and limonene. The other half of the participants slept in rooms in which only water was nebulized as a placebo treatment. The scientists drew blood from each person before they went to sleep and then again at seven o'clock in the morning.

The results were groundbreaking. The participants in the "tree group," who had breathed tree terpenes at night, showed a clear spike in the number and activity of the natural killer cells as well as an increased amount of anticancer proteins in the blood.[13] The positive impact increased from night to night. This means that the same effects occurred in the hotel as had in the forest. The effect is much stronger in the forest, however, because we breathe a rich mixture of terpenes and other bioactive substances there. The effect of isolated terpenes in the hotel was nevertheless significant. The participants whose room air was not enriched showed no changes in their blood.

This experiment was an ideal double-blind study. The researchers managed to keep their test subjects completely in the dark about the fact that they were inhaling terpenes at night. They had no idea what the experiment was about: placebo effect eliminated! Furthermore, the second group received pure water vapor through the same type of device as a placebo and showed no health effect. The doctors who drew the blood samples in the morning did not know who was being treated with hidden terpenes and who was not. Besides, the measuring instruments in the laboratory with which blood values are analyzed are 100-percent objective. This ingenious experimental design shows that it is possible to research nature-human medicine with the utmost precision. The scientists turned hotel rooms into perfect labs.

The next piece of the puzzle was another test of the terpenes' effects on human killer cells: Parallel to the double-blind study in the hotel, the forest scientists treated human killer cell cultures in the lab with various terpenes from forest trees. Lo and behold, even in the petri dishes, these tree substances caused the natural killer cells to become more active and the anticancer protein content inside to increase.[14]

It has therefore been made clear that the health-promoting effects of the trees are linked to terpenes that plants emit during biochemical interaction and communication with other living organisms.[15] Numerous studies have proved this according to the gold standards of medical research. Scientists from entirely different fields of research also backed the findings of the forest scientists from several angles.

The cancer researcher Roslin Thoppil, from Vanderbilt University in Nashville, and Anupam Bishayee, professor at the Larkin University College of Pharmacy in Miami, discovered the effects of terpenes against malignant tumors completely independent of forest medicine research. They analyzed numerous laboratory experiments from different research centers around the world and compiled the results. It turned out that cancer researchers had already demonstrated the effectiveness of plant terpenes against tumors in cell cultures several times over.[16] Laboratory experiments also confirmed the fact that terpenes strengthen the natural killer cells and trigger increased production.[17] International cancer researchers have long strived to use terpenes in the medical treatment of tumors. Most of these scientists are not directly interested in trees. Their results, however, support the findings of forest medicine. Exploring the healing powers of nature is a highly interdisciplinary project.

There is another important overlap. Both cancer research and forest medicine identified the same terpenes as the most effective, namely pinene and limonene.[18] Both types of terpenes are found in trees and other plants. One place in which pinene forms is in the green parts of the conifers, for example in spruce, fir, and pine needles, but it also occurs in Mediterranean myrtle shrubs. Citrus and lavender plants contain limonene, and spruces, firs, and pines emit limonene as well. Both terpenes are also found in high concentrations in the essential oils of these trees. Forest medicine researchers also came to the conclusion that conifers and evergreen plants yield the highest amount of and most effective terpenes. This does not mean, however, that deciduous trees, such as oaks, beeches, or maples, don't form health-promoting terpenes as well. All terpenes from trees contain a substance called *isoprene*, which has proved to have a stimulating effect on our immune system as well.

This concordance between the mentioned studies is of great importance and opens up new and promising paths in cancer therapy. Cancer researchers discovered in laboratory tests that limonene caused 80 percent of breast carcinomas to shrink or disappear.[19] The effectiveness of different terpenes in cancer has been demonstrated again and again. Other terpenes from plants act, for example, against skin cancer, kidney and liver cancer, and other tumors.[20] They cause the tumor cells to die and inhibit their activity. Scientists are currently intensively pursuing the effects of terpenes, as they are a potential cure for cancer in the future.

We find the highest concentration of terpenes in the forest air during summer, when plants are biologically very active. In winter the concentration is lowest, but not zero. The concentration is higher in the middle of a forest than on the edge, since the tree crowns are closer together and contain gaseous substances better. After a rain and during fog, the air is particularly rich in terpenes.

Based on his studies in recent years, Qing Li specifically recommends spending two full days a month in a forest area, staying outdoors for as long as possible. In this case, the effect on the immune system lasts one month before it needs to be refreshed. In order to achieve this effect, it is not necessary to exercise or hike in the forest. We absorb terpenes mainly through the lungs, but also through the skin—in other words, by simply being present. Substances from trees enter our blood circulation, reaching our organs and cells this way. The limbic system in our brain also decodes and responds to terpenes by releasing neurotransmitters and hormones that benefit our health.[21] We will deal with this in more detail later. Qing Li described forest medicine as "natural aromatherapy" and "rest and relaxation, while inhaling gaseous organic substances."[22]

People who have read my earlier books sometimes ask me if I hug trees. No, I do not. I don't have anything against hugging trees, but it's simply not me. It could be, however, that hugging trees does have an especially positive effect on our health. Tree bark is actually one of the richest sources of terpenes. When we hug a tree, we move very close to this source with our respiratory organs and skin. Perhaps people who hug trees intuitively feel that there is *something* good for them there. In the middle of winter, when deciduous trees no longer carry leaves, the gaseous terpenes flow almost exclusively from the bark and from the roots underground. So far, nobody has researched whether hugging a tree makes our immune system react by strengthening its antibodies. This question would make for an interesting study by biologists, biochemists, or medical researchers.

Tree hugging could also affect our health through neurobiological mechanisms. In a later chapter, I will address how the human brain reacts in a similar way to an encounter with a loved one as it does to an encounter with animals and plants. It pours out "binding" hormones, which until now have only been explored in the context of interpersonal relations. These substances lead to several positive effects in the body. It was not until recently that science discovered that these special encountering and binding reactions among humans function the same way among different species at a neurobiological level and trigger a comparable effect. In addition, contact with trees activates the parasympathetic nervous system ("rest and digest"), which has a very positive effect on our organs. I'll come back to that later, as well.

## Change of Perspective

"We are faced with the surprising fact that the immune system is a sensory system that is capable of perceiving, communicating, and taking action," wrote Joel Dimsdale, a professor of psychiatry at the University of California in San Diego.[23] He was referring to the latest findings in immunology, which suggest that the human immune system—and, of course, that of animals and plants—is in a much higher degree of exchange with the environment than was previously known. However, our "sensory organ immune system" does not only communicate with its environment. It also has highly complex communication with the hormonal system, nervous system, and other organs of the body. It has even been demonstrated that our immune system communicates with the gut bacteria living within us—the "good" intestinal bacteria that strengthen our defenses. Through modern procedures, it is already possible to follow how our mental processes affect the immune system and organs. I'll explain in detail later how our body succeeds in translating not only substances, but also sensory stimuli of nature into the language of our organs.

It is interesting that our own immune system responds to some of the terpenes in the forest air in a similar way as the trees themselves do, namely with an increase in defenses. To conclude that our communicating immune system "understands" the chemical words of the trees (that is, that it communicates directly with the trees) would be a far stretch. What is certain, however, is that our immune system knows how to decode the terpenes in the forest air and reacts to them in a predictable way. We can safely say that tree terpenes are nothing new to our immune system and that it knows what to do with them. These are healing codes from nature, which isn't a surprise. What would be surprising, however, is if the human immune system did *not* know what to do with forest terpenes and other substances in nature.

Our immune system developed over incredibly long periods of time. It *evolved*, as we say in the field of biology. Under the term "evolution," life scientists understand a continuing process of living organisms adapting to their environment. Evolution is change. Genetic traits are also changing, which are passed down to the offspring during reproduction. Thus, every living thing—whether human, plant, fungus, animal, or single-cell organism—carries the entire past of its species within. Formulated for the first time in the nineteenth century by Charles Darwin, the biological fact that living beings are evolutionarily changing and continually adapting to their environment provides a plausible explanation for the enormous biodiversity of life. This also means that we are in a position to help shape the future biological development of the human species through our collective behavior and the shaping of our environment.

Representatives of our species, *Homo sapiens*, have existed for at least 100,000 years, and according to the latest findings, perhaps even for 400,000 years. In Israel, fossil teeth from this age were found in a prehistoric cave. Human genetic studies have revealed that early *Homo sapiens* from Eastern Africa emerged from a relatively small population of ten thousand people of the genus *Homo erectus*. Of course, the human immune system did not wait to develop until the transition between *Homo erectus* and *Homo sapiens*, but rather developed continuously. It can be traced back to our early ancestors and far into the history of mammals. Besides our closest relatives, the chimpanzees, many other mammals have immune systems similar to ours, even though they branched off from our phylogenetic tree millions of years ago.

Our ancestors from the genus *Homo* and *Australopithecus* alone create a history of four million years on earth. Our immune system developed in the ancient network of life, with which it is still connected today. This network also includes plants and, therefore, trees. Our immune system evolved *from* nature, *in* nature, and closely *with* nature. As a communicative sensory system, the immune system goes back an incredibly long way in its biochemical exchange with trees. In simple terms, our immune system has breathed tree terpenes for ages, and one could assume that its exchange with bioactive substances in nature is necessary for its holistic functioning. We are *functionally* connected to nature and, therefore, based on biochemical reasons alone, inseparable. How this came to be will become clearer in the course of this book.

We are still a part of the network of life. This fact alone is reason enough to change our perspectives. Instead of saying contact with tree terpenes improves our health and increases natural killer cells and anti-cancer proteins in our body, we should ask ourselves whether it is not the other way around: *separation* from natural ecosystems of which we are a part (and thus separation from tree terpenes) ends up causing us to carry *fewer* natural killer cells and *fewer* cancer-protective substances than we would in natural conditions. Not only do environmental toxins and negative influences of modern life make us sick, but so does the absence of bioactive substances from the plant world that our body needs to stay healthy.

This is the only way to interpret the scientific findings. Evolution is always a process of adaptation to environmental conditions, and our natural habitat, to which we have adapted, does not consist of concrete and asphalt, but of soil, trees, shrubs, grass, ferns, fungi, lichens, microorganisms, and animals. It is an organic habitat full of bioactive botanicals. This is where our immune system is at home. Separating ourselves from these substances leads to a weaker immune system when compared to being in our natural environment.

I'd like to go back to the studies that I mentioned at the beginning of this chapter. Scientists from different universities around the world—physicians, biologists, psychologists—repeatedly encountered the correlation between trees and lifestyle diseases. More trees mean a lower risk of cardiovascular disease, diabetes, metabolic disorders, hypertension, and even cancer. The prevalence of these lifestyle diseases is therefore not only due to the damaging influences we are exposed to, but also to the deprivation of nature's substances. I will continue to underline this perspective in the course of this book, and we will see that separation from nature is not only responsible for many physical diseases of our time, but also for the prevalence of mental illness.

Another conclusion can be drawn from the previous findings: the human organism does not end at the external borders of the body. We will be dealing more closely with this assertion as well.

## THIS CHAPTER IN A NUTSHELL

International studies show a strong link between the presence of trees and human health. Renowned scientific journals such as *Science* and *Nature* have already published such evidence. The mere view of a tree from a hospital window activates self-healing powers in patients after an operation. The danger of suffering from modern lifestyle diseases diminishes with the increasing number of trees around a person. According to one Canadian study, the health effect of ten additional trees around residential buildings in the city of Toronto would be equivalent to the individuals living there being seven years younger. To get to the bottom of these immensely positive effects on humans, scientists all over the world have looked for explanations.

Among other things, they discovered that trees emit chemical "words" in plant language. These are molecules from the class of compounds called terpenes, which bear meanings in the plant world, similar to the vocabulary of human speech. Our immune system, which according to the latest findings is a sensory system capable of communicating, reacts to these substances with a significant increase in defenses and mechanisms protecting against cancer.

In numerous laboratory experiments, scientists verified that these effects come from tree terpenes. International cancer researchers, who otherwise have nothing to do with trees, discovered in laboratory experiments that terpenes from trees act against tumor cells. These scientists even underlined the pharmaceutical potential of plant terpenes as highly concentrated chemotherapeutics in the future treatment of cancer.

Because frequent contact with nature has such a positive effect on human health, I think it's time for a change in perspective. Our immune system is not *strengthened* by substances from trees, but is rather *weakened* by the separation from these substances in modern life. Spending time in nature therefore does not lead to "more" defense cells; rather it brings their number and activity back to a natural level. This can also be well understood from an evolutionary angle because we evolved in nature as part of the network of life. The bioactive substances of the trees are not new to our immune system; rather they have integrated themselves into its functioning over the course of evolution. We cannot blame the prevalence of lifestyle diseases only on environmental toxins, but also on our separation from nature.

# 2

# EVOLUTION AND MEDICINE

## The Story of My Knee Joint

Several years ago, I was suffering from pain in my left knee. With every step, I felt a very unpleasant snapping in my knee joint. It felt like there was a rubber band popping back and forth behind my kneecap. In addition, there was an inflammatory pain and a soft grinding whenever I stood up from a squat.

To get to the root of these issues, my knee landed in the tube of a nuclear magnetic resonance scanner, also known as magnetic resonance imaging or MRI. With loud rattling and bass-like knocking, the device radiated a strong magnetic field as well as electromagnetic radio waves through the joint capsule to scan the internal structures of my body—in this case, my knee. Since atoms in the human body respond to those magnetic fields by emitting electromagnetic energy, the device could create an image of my knee from these resonances (hence the name, magnetic *resonance* imaging). The doctor interpreted the image and came to the conclusion that there was a tear in my meniscus. He claimed surgery would be unavoidable in order to restore the full function of my knee.

With the images from the MRI, I went to a renowned knee specialist to get a second opinion. He also recommended surgery as soon as possible. Without surgery, I was assured that my knee joint would completely lock up. That kind of thing could occur at any moment. Who wants to bite the dust while crossing traffic or hiking because their knee suddenly locks? I made an appointment with the doctor for the operation.

What happened in the hospital could have been material for a comedy. I was taken to my room after arriving early in the morning. The anesthetist gave me the usual spiel while I sat in front of her in a loosely fitting white gown. The surgeon was preparing to receive me in the operating room. Everyone was ready to go. An assistant physician smeared shaving cream on my knee and carefully began shaving off the hair there. I looked down at my naked knee. It wasn't until that moment that the ugly truth showed its head, and it became crystal clear to me what was about to happen. After the shave, the surgeon was going to drill at least two holes into the capsule of my knee joint in order to squeeze instruments under the kneecap and chop and sew my meniscus. I saw my untouched knee and imagined what it would look like in a few minutes. Something inside me resisted. I didn't want to make a drilled mess out of my otherwise healthy knee.

"There won't be any knee operation today," I said. "At least not mine." A pair of eyes looked up at me, and the shave came to a sudden stop. More pairs of eyes—this time, the nurses—looked at me equally astounded. I had interrupted their surgical routine, and clearly no one had counted on it. The assistant doctor looked up and said he needed to get the surgeon. I didn't care one bit that the entire surgical team was already there prepping. I jumped up, took off my white gown, and packed up my stuff.

The surgeon came in and assured me one more time that my knee would horribly lock up if I didn't have the surgery. These types of words carry a punch. They create fear and insecurity, especially when they come from the mouth of one of the most renowned knee experts of Austria, my homeland. Nevertheless, I signed a form declaring that I was leaving the hospital of my own free will. Otherwise, I was sure that I would leave that hospital with two holes in my knee, several packets of painkillers in my bag, and the risk of postoperative complications.

And there was not a single tree around that could have had some counter-effect. The supposed knee lockup, on the other hand, still existed only in my imagination. That is how I ended up walking out of the hospital door that same day with my suitcase in hand—popping knee and all. I felt so relieved.

Shortly thereafter, I was lying in yet another orthopedic surgeon's examination room. He examined my knee with his experienced hands, moved my knee joint, and then folded his hands on his lap. "Your meniscus is fine."

I pointed out the MRI results, yet my new physician was not interested. He preferred to rely on the good, old-fashioned method of using his hands for a physical examination, which has long been an essential component of diagnosis in orthopedics. Using this method, he said that an experienced orthopedist can accurately determine the condition of a meniscus.

This tried-and-true method indicated that I had a so-called discoid meniscus. A look at the MRI images confirmed his suspicion. My new practitioner interpreted the MRI completely differently than his colleagues. The meniscus is usually shaped like a horseshoe. The area in the middle remains free so that the cartilage of the knee joint can move in it. For me and some others, the meniscus is more like a disk. The recess in the middle is not as pronounced as it is in the average person. This results in friction on the meniscus when the knee joint is moved. The meniscus often gets pinched for a moment, which sounds like a snap in the knee joint. Over the years, the meniscus can also fray a little in the middle, and this is often misinterpreted as a tear in imaging diagnostics. A discoid meniscus is not a disease but rather an anatomic variant of nature. Although it is often treated surgically, an operation would only be necessary in rare cases—mainly for overweight persons, which I am not. My new doctor recommended bike riding.

"I just have to ride my bike?" I asked.

"Yes, bike riding on a regular basis strengthens the knee muscles and solves the problem," the orthopedist replied.

I followed his advice, and within two months all symptoms in my knee had disappeared. I felt no more snapping, no grinding, no popping, and no pain. Without medical measures and without surgery, my problem was solved. In addition, I discovered a new passion that has been part of my life since then—mountain biking.

Cycling is a particularly knee-friendly sport due to the circular movements. Especially when riding against medium resistance (that is, when you exert yourself without standing on the pedals), bike riding strengthens the muscles that run along the entire knee. Since it is the muscles that maintain our body upright and hold our cartilage and soft parts together like a corset, strong muscles are also able to stabilize the knee joint from the inside out.

The pressure on the knee cartilage and the meniscus—in my case, on the discoid meniscus—decreased, and the symptoms disappeared. Since then, I have never had any problems with my knee, and I bet I will never suffer a knee locking up, which my highly esteemed doctors had predicted, actually "guaranteed." If, on the other hand, I had undergone the surgery, it could have resulted in damage to the knee joint, and the surgical perforation of the joint capsule can also lead to complications.

In December 2013, the *New England Journal of Medicine* published the results of a clinical trial from a Finnish meniscus research group. A large team of medical scientists and physicians conducted a study in which they randomly divided patients who had been diagnosed with a torn meniscus and were suffering from related symptoms into two groups. Both groups received an arthroscopy of the knee under local anesthesia, in which a camera was inserted into the knee joint to diagnose the knee problem. In the first group, the doctors also performed a routine meniscus repair. This was done through the holes that had already been made during the arthroscopy. During the operation, meniscus tissue was partially removed, a standard procedure that is often carried out.

The second group received only a "placebo surgery" after the joint examination. The physicians tricked these patients by showing them a recording of another person's knee surgery, while letting them believe they were observing their own meniscus operation.

As is typical after these kinds of interventions, all patients had to undergo physical therapy for weeks or months. Scientists observed the healing process for an entire year. It was important that the researchers who documented the progress were not the same as the ones who operated. They did not know which patient actually had the surgery and which only appeared to have had it. This was a classic double-blind study.

The results were astonishing. The patients who actually had surgery did not achieve any better results than the patients with the pseudosurgery. It turned out that for successful healing it was completely irrelevant whether the surgical procedure took place or not.[1]

The researchers partially attributed the success of the pseudosurgery to the placebo effect (that is, the effect of fake treatment through mental mechanisms). This phenomenon is a fascinating reference to the connections between physical health and mental condition. The placebo effect has been neglected in the past by many scientists and is now increasingly becoming the focus of medical research. However, in the case of the Finnish study with feigned operations, they assumed that the recovery of patients, with or without surgery, was mainly due to the physical therapy that all of the patients in the study received.

Some scientists have long criticized the fact that there is no clear proof of the effectiveness of partial removal of the meniscus in cases of knee problems. They believe the surgery is routinely carried out far too often. Studies have even led to the conclusion that joint inflammation may be one of the late secondary results of meniscus surgeries. These negative consequences often occur fifteen or twenty years after surgery.[2] There are certainly cases where partial surgical removal of the meniscus is unavoidable, but it is clear that this surgery should be placed on the list of too-often-performed operations (of which there are several more).

The Finnish meniscus study demonstrates what I can confirm from my own experience: natural movement and muscle building are the way to healthy knees and help stabilize already damaged knee joints. In my case, mountain biking and jogging through the forest were my salvation, and for participants in the mentioned study, physical therapy. The human knee needs to be treated in a "knee-friendly" manner, and asphalt and concrete are not the right environment for this. The hard pounding causes damage to joints and cartilage tissue. Special orthotic inserts for shoes, of which the market offers plenty, do not help much. Most people move too little and sit too long. The modern knee receives too few knee-friendly movements and remains immobile far too many hours every day while we're sitting.

My own knee problems started in a phase of my life when I had little time for trips into nature or sports. I worked at the time for a company that expected everyone to be in the office fifty to sixty hours a week. On top of that, I was taking a continuing education course. I led a life between desk, subway, and university library.

The correlation between the modern urban and professional life and the condition of the human musculoskeletal system is probably obvious to almost every human being, and with my readers, I am certainly preaching to the choir. Environmentalists have gathered convincing evidence that the human body leads a life in this modern world that is not at all suited to the species and thus damages it. However, the implementation of these findings is lacking in medical practice—and is entirely missing in the shaping of our lives.

In order to treat the human knee and take care of it in a knee-friendly way, we should not be quick about drilling holes into our joints and entirely or partially removing body parts. Instead, we need to fully understand how the knee joint functions and take this into account in daily life and during therapy. For this, we must take a look at the evolution of the knee and its natural environment.

## Evolution of Disease and Health

When I began to treat my knee in knee-friendly ways again, my issues disappeared. Along with mountain biking, hiking and running on natural terrain helped. The soft forest soil was a pleasure under my feet while jogging. The unevenness of meadows, forest paths, and rocks, which my muscles had to balance out constantly, even trained the smallest muscles of my knee and built up strength and stability from the inside. A knee-friendly life in a knee-friendly environment was the best therapy. The natural environment for our knees—just as for our immune system—is the world of plants and animals, fungi and forests, and meadows and rocks.

"Nothing in biology makes sense except in the light of evolution."[3] These words came from the geneticist and evolutionary biologist Theodosius Dobzhansky, who was born in 1900 in Nemyriv (then part of the Russian czarist empire and now in Ukraine) and spent his life as a scientist in the United States. Dobzhansky shaped modern evolutionary biology by associating Gregor Mendel's laws of heredity with the doctrine of evolution. When he said that nothing in biology makes sense except in the light of evolution, then he must have meant illnesses too.

Taking into account the evolution of our species is a future-oriented approach in medicine. At universities, biologists and medical scientists therefore work together on what's known as evolutionary medicine. In the everyday practice of most physicians, evolutionary thinking has not yet played any role. But this will surely change in the coming decades because evolutionary medicine gives us the chance to understand the emergence of diseases in a much larger context and thus to also understand it *better*. This is about more than just trivial insights such as the fact that walking on asphalt does not benefit the knee joints of humans coming from virgin forests and savannas.

Let's recall the evolutionary age of the human race. *Homo erectus* emerged on the world stage almost two million years ago, and our own species, *Homo sapiens*, has existed on earth for 100,000 to 400,000 years. The most modern subspecies *Homo sapiens sapiens*, to which you and I belong, has been around for 40,000 years. Most of the diseases we are dealing with today arose at the earliest 10,000 years ago or much later. Some, such as AIDS, are only a few decades old.

More than half of the infectious viruses and bacteria to which we are exposed developed after our ancestors had transitioned to farming and started breeding animals. The pathogens originally passed from farm animals and pets to humans. When the term "swine flu" circulated in the international media in 2009 and 2010 and caused an uproar all over the world, few people probably knew that even ordinary influenza viruses were once transmitted by pigs to our ancestors.

From this point of view, every flu has always been a "swine flu." Smallpox could be called the "camel virus" according to this logic because it can be traced back to camels. Following the same thought pattern, measles would be called the "bovine disease," and herpes might be called "chimpanzee fever."

In the early 1980s, when the human immunodeficiency virus, HIV, which causes AIDS, was discovered, scientists found a relative of this pathogen first in domestic sheep. It soon became apparent that among the primates (besides humans), the macaques also suffered from symptoms that were very similar to AIDS. Today, molecular biologists assume that the human HIV virus is derived from a nonhuman primate virus in central and western Africa. It is called SIV, the simian immunodeficiency virus, or the "monkey immunodeficiency virus." Over the past decades, HIV has mutated several times, making the search for a cure very difficult. Because of its ability to adapt rapidly, the pathogen eludes medicine too quickly.[4] If we better understand the evolution of the HIV virus one day, we will be able to find effective remedies for this pathogen. Evolutionary medicine will most likely be the key to the cure for AIDS.

## Does Our Separation from Nature Make Us Sick?

Whether someone is near- or farsighted or has perfect vision depends on the length of their eyeball. A too short eyeball leads to farsightedness, too long to nearsightedness (myopia). In industrialized countries, myopia is growing rapidly. This increase cannot be explained by heredity alone. As studies have clearly shown, nearsightedness is linked to the fact that many of us do not spend enough time outdoors.

In order to enable correct vision, the human eye gradually develops its final form over the course of childhood. This process can take up to ten years, after which no significant adjustments are expected. In the first years of life, almost all children are farsighted because the eyeball is still too short from the front to the back. However, in most cases, the farsightedness is compensated for by the lens, which is more elastic in children than in adults. The eye gradually achieves correct vision while the child grows. The eyeball becomes longer until the eye sees normally. It slowly forms by interacting with its environment. How the human body exactly accomplishes this impressive feat is still unclear.

Nearsightedness occurs when the eyeball grows too long. From clinical studies, we know that natural daylight inhibits excessive growth of the eyeball and thus protects us against myopia.[5] This confirms the rule that we have already established for the immune system and knee. The human body requires the influence of its natural habitat (that is, the environment in which it evolved over eons) to "function" correctly.

One of the reasons for the rapid increase in myopia is, in fact, that we adults, as well as children, spend too little time outdoors these days and increasingly more time in artificially lit rooms. This is not a hypothesis, but an occurrence that is supported by the fact that myopia is more frequent in cities than in rural regions. In large metropolitan areas like Tokyo, nearly 90 percent of the residents are nearsighted. The average for industrial nations is about 35 percent worldwide, and some studies report an even higher percentage of myopia.

The lack of natural sunlight is not the only culprit for the increase in nearsightedness. Children and adults are also performing more daily tasks where their eyes need to focus close up. We are in front of the computer too much, where our eyes have to work overtime due to using a monitor that is too close. For this reason, ophthalmologists recommend that you frequently look at distant objects while working at a computer. Trees are perfect for this, since the sight of them triggers healing effects, as we already know.

Since the eye slowly gains correct vision by growing in length in the first ten years, children should sit in front of the computer as little as possible and play more outdoors where daylight reaches their eyes and where they must always focus at different distances. This also trains the eye during the growth phase and gives it the opportunity to constantly adapt its growth through interaction with the environment. Experiences in nature are therefore an important factor for the development of the eyes of our children. There has been a shift, however, in childhood activities from outside to inside, and the use of computers is encouraged and required in schools earlier than ever. This is clearly a negative development that contradicts the findings of evolutionary medicine.

The fact that our eyes need to go through a process of adaptation in childhood has evolutionary causes. In the course of the development of life on earth, eyes in the animal kingdom arose many times. Invertebrates in the sea, such as the octopus and squid, have eyes constructed similarly to our own, with a lens, cornea, and vitreous body. However, their light receptors are facing the light. This is a better construction than with humans and all the other vertebrates. Our receptors lie, in simple terms, on the back of the retina, on the side facing away from the light. The light must pass through the blood vessels and nerve paths of the retina. These throw shadows, which our brain has to remove so that a clear image emerges. Our eyes are therefore, in a sense, the "wrong way round." Moreover, the better designed eye of the octopus and squid has evolved from a cavity in the skin, while our vertebrate eye formed from a bulge of the cerebrum. This means the eye is a part of our central nervous system. The special design and inverted position of the light receptors makes a phase of adaptation necessary in the first years of life.

Another medical example of how our evolutionary past becomes medically relevant is obesity. Have you ever wondered why sweet carbohydrates like sugar are so popular and why so many people like eating fatty foods? Sugar and fat seem to trigger in us an almost uncontrollable urge for more. A common and enlightening explanation from evolutionary biology states that high-energy food was scarce for our ancestors in the Stone Age. It was a question of survival to develop a taste for it and to actively seek that taste experience of high-energy sustenance. Sugar and fat contain particularly large amounts of energy.

Today, we live in the age of excess, in which the food industry pursues its commercial interests. It deliberately exploits the archaic patterns of consumers in order to make profits. Especially refined sugar and fat can trigger a kind of addiction. Our "Stone Age body" is simply not made for industrial food. Our ancestors' useful predilection for high-energy nutrition is completely out of place in the modern age. The consequences are excess weight, fat accumulation around the stomach, high blood pressure, disorders of fat metabolism, as well as type 2 diabetes. These lifestyle diseases, which physicians summarize as "diseases of affluence," are indubitably due to eating habits that are far removed from the realities to which the human body had formerly adapted. In addition, lack of movement is a further risk factor. Our body is evolutionarily designed for regular movement, not for hours of sitting and standing. Diseases of affluence are regarded in the medical field as serious risk factors for blocked arteries and heart attacks.

Studies of nature-human medicine have shown that even light activity and walks in the forest lead to a significant reduction in blood pressure and a lowering of blood sugar levels for people with type 2 diabetes. Inhaling tree terpenes promotes the formation of endogenous substances that protect the heart. These cases point out once again that, in addition to the damaging influences and habits that affect us nowadays, the absence of regular contact with nature also plays a role in the development of diseases.[6]

## Cancer Began with LUCA

Since Theodosius Dobzhansky was obviously right about nothing making sense in biology except in the context of evolution, and because this insight also applies to diseases, the question arises as to what explanations evolutionary medicine has to offer for diseases such as cancer. How can the threatening mutation of cells into tumors be understood in relation to evolutionary mechanisms?

Cancer existed long before our time. We know this because of a find by paleobiologists, scientists who focus on extinct species of bygone ages. They dug up a fossil of dinosaur bone with a petrified tumor that was 150 million years old. This means there must have been cancer at that time, although it is impossible to say how frequently this disease occurred.

The potential to grow a tumor has probably existed in cells since the beginning of life. Let us suppose that life on earth, as most biologists assume, arose four billion years ago. At that time, the first living creatures appeared on our planet. So far, there are only unproven hypotheses and wild speculations about their origins, but it is highly probable that they were unicellular organisms living in the water. Biologists call the earliest of all life-forms LUCA. This acronym stands for "last universal common ancestor"—a sexless, unicellular ancestor. For the sake of our discussion, let us forget for a moment that some biologists think it is possible that LUCA was transported to Earth by a meteorite from another part of the universe. This would not change the fact that all living things on earth—trees, shrubs, and the other plants, fungi, bacteria, as well as animals and humans—are distantly related to LUCA through billions of years. That means that a little bit of this ancestor is in all of us.

This prehistoric cell, which was at the beginning of the evolution of life, must have already had genetic material and been able to propagate (that is, to reproduce). Single cells have two ways to multiply. One method is to fuse with other cells. This is a simple and very old form of sex, which works according to the same principle as with us humans. On the other hand, unicellular organisms can simply split in order to multiply. They duplicate themselves, getting "two for the price of one." They do not need another cell. This is the asexual way, the most ancient form of reproduction, of which LUCA must have been capable. The offspring that are produced are genetically identical to the cell from which they emerged. They are carbon copies.

We humans, in the entirety of our cells, are descendants of LUCA. Consequently, it is not surprising that our body cells employ essentially the same asexual form of duplication that we know from single cells. They divide and create new cells from within themselves during the whole lifespan of a person. The offspring of our cells are usually copies of their mother cells—clones—that exactly fit their purpose. A liver cell must emerge from a liver cell, and a skin cell from a skin cell. In this way, our body is continually renewed by cell division. The epidermis, which is the outermost layer of the skin, completely renews itself every four weeks. The cells of the liver are much slower and require 140 to 400 days to renew themselves. Some cells do not regenerate in adults at all—for example, some types of nerve cells.

If body cells regularly regenerate, the old cells must die to make space for the new ones. For this reason, body cells voluntarily die after a certain time. Because this voluntary death of the cell is genetically coded, we refer to it as "programmed cell death." And at this point, things becomes complicated because the important programmed cell death contradicts the primal interests of each cell, which is to exist and evolve.

The evolution of life is based on changes in the genetic program. A tiny "photocopier" in living cells makes mistakes when reading and writing this program. This results in mutations—that is, changes in the genome of the cell. These mutations are *one* of the reasons evolution continues at all. If the original cell LUCA, from which we all descended, had always produced perfect and error-free copies of itself, there would be no other form of life on earth today. There would probably be no life at all, since the environmental conditions on earth have continued to change dramatically in four billion years, and the LUCA form of life could not have adapted to its transforming habitat without changing. Mutations, transformation—these are the engines of evolution that are at work in every cell.

A body cell can become immortal by mutation. Programmed cell death no longer takes place, and the cell continues to divide rapidly. Since it passes on its altered genetic material, it produces new cells that are also immortal. The mutated, immortal body cells accumulate, instead of making room for new cells. A tumor develops, which spreads and can subsequently damage and destroy organs of the body. Mutation is nothing more than an evolutionary process of the cells that are blind to the human condition to which they belong. Consequently, research in evolutionary medicine is of great importance for the further development of cancer treatment.

Every day, damage occurs thousands of times to the genetic material of our body cells. For example, free radicals, which form during chemical processes in the human body, attack the genetic material that is in our DNA. Free radicals are extremely reactive compounds that whiz like crazy Frisbees around our body in search of a chemical binding partner. They can even rip out components from the DNA of our cells.

The damage they cause can also lead to the formation of dangerous cells. We, therefore, have repair mechanisms that restore the damaged genetic material. If the repair fails, the dangerous cells can also spread.

It is a constant balancing act to protect our body from cancer, because cancer stems from the primal survival instincts of life. In order to prevent a tumor from growing, the body's defenses must be permanently alert and prevent the basic life energies of the cells from derailing and making us sick. For this reason, doctors and biologists speak of a "cancer lottery."[7] In addition to genetic factors, the environment and immune system play an important role in this cancer lottery. Environmental contaminants can also damage our DNA. The repair mechanisms can no longer keep up with all the work, and the risk of cancer increases. Equally harmful is an unbalanced diet, which can promote the development of free radicals. For this reason, nutritionists recommend an abundant intake of secondary phytochemicals from vegetables and fruit. Some plant substances are able to inhibit the formation of free radicals in the body. Can you guess which ones? Almost all cancer-preventative substances from fruits and vegetables are part of the group of terpenes! We already know this substance group from the forest air. One example is the carotenoids. These are terpenes that are common in yellow and red fruits and vegetables, such as carrots, tomatoes, pumpkins, red beets, and peaches, but also in broccoli, spinach, and more.

Our immune system is an especially important factor in the fight against cancer. I have already described how natural killer cells and anticancer proteins detect dangerous cells and force them to commit suicide and that all these defenses can be activated and multiplied by inhaling tree terpenes. By no means is the state of the natural environment of Earth inconsequential to the state of our health. The relationship between humans and nature is decisive for the outcome of the cancer lottery. Functioning, clean ecosystems, of course, have fewer carcinogenic effects than a polluted planet, and we can inhale more health-promoting plant substances like the terpenes. The influences of nature let us play the cancer lottery with better odds because they strengthen the anticarcinogenic mechanisms of the body and help the immune system to function properly. In terms of evolutionary biology, we can say that alienation from nature increases the fitness of potential cancer cells. On the other hand, the effects of ecologically intact habitats strengthen the fitness of our defense cells and other natural antagonists.

Although cancer existed long before *Homo sapiens*, it has been one of the most frequent causes of death in humans since the twentieth century. This is not the case in the animal kingdom. Cancer is a human lifestyle disease of the modern world. The mechanisms that enable cancer cells to become resistant to chemotherapeutic agents are also evolutionary. Similar to antibiotics in the case of bacteria, chemotherapeutic agents increase the artificial selection pressure on tumor cells. Doctors use poisons to kill tumor cells. However, cells have tiny pumps they can use to pump out toxins. Some tumor cells are fast and efficient enough in pumping out toxins to survive chemotherapy. That means these cells become resistant to it. When almost all cancer cells except the resistant ones are killed by a chemotherapeutic agent, even more severe tumors can emerge that contain nothing but copies of the resistant cells. The treatment of these "super tumors" usually has a greatly reduced chance of success. Therefore, cancer treatment of the future must include evolutionary findings and discover evolution-proof forms of treatment. To this end, network thinking is necessary. Humankind needs to be seen for what it is—namely, a natural species that emerged from the network of life and that can never, under any circumstances, remove itself from this context. A medicine that ignores the nature-human relationship will never be a successful medicine.

## Evolution Glasses: Path to Knowledge or Misguidance?

It is in fashion to explain characteristics of people through evolution. This trend goes far beyond the human body and the emergence and treatment of cancer. Numerous bestselling authors have used evolutionary justification for almost every detail of human culture and coexistence in the past few years. Women are this way and men that way because they did this or that during the Stone Age. Some psychologists do not hold back with their dazzling imagination when interpreting human behavior.

When human behavior is explained by the evolutionary past of our species, sociologists often object. They believe humans are a learning creature. Behavior is therefore learned and not biologically determined. During a 2008 colloquium of the German foundation Bündnis Mensch & Tier (Alliance of Humans and Animals), sociology professor Helga Milz from the University of Hamburg, behavioral biology professor Kurt Kotrschal from the University of Vienna, and other scientists talked about these perspectives.[8] In the course of the discussion, Kotrschal, the behavioral biologist, said that culture without biology was not possible and that the biology of mankind is the *basis* for our culture.

The sociologist Milz then replied, "Yes, there is no culture without living people, so you are right. But they have no innate behaviors—that's all learned."

Kotrschal's short answer was, "Everyone is stuck in their norms of reaction."

Milz thus assumes that the only biological prerequisite for culture is the fact that humans are living, and therefore biological, beings. However, she insists that human behavior is exclusively learned and not biologically assigned. Kotrschal counters that every human being moves within a "reaction norm." This means that every person—and by the way, every animal—adopts a behavior by learning it, but this is only possible within a biological framework. Biological surroundings are the prerequisite for a certain behavioral repertoire of a person. And this was also shaped by the evolutionary history of humankind.

The discourse between the two worldviews of sociology and biology continued. Anke Prothmann, a pediatrician at the University Hospital of Munich, commented, "There is a study conducted with three-month-old infants that examines the question of which moving stimuli infants prefer. And at the age of three months, they were able to recognize biological movements and find them more attractive. How is that learned?"

The sociologist Milz answered, "It is vital for survival. They had already learned that a long time ago."

The pediatrician then referred to her own experiences as a mother. There was no doubt that her son, shortly after his birth, could distinguish between a maternal figure and, for example, a dog. Is this not proof that it is "programmed" in newborns to recognize their human mother and to distinguish her from other living beings? After all, this distinction was extremely important for the survival of newborn babies in the Stone Age.

The sociologist Milz repeated her argument, "Yes, but that does not mean anything. It is all learned." Even basic survival functions like recognizing one's own mother, as far as the sociologist is concerned, are not biologically determined. The quintessence of her assertion is that people have no instincts, and only animals have such a thing.

The behavioral biologist Kotrschal brought the dispute diplomatically to an end and pointed out that the disagreement of nurture versus nature in science has been off the table for forty years. "You can only learn what you have a brain for," he said. "Innate is not the opposite of learned."

It is undisputed that human behavior and culture have a biological *foundation*. The extent of what we learn and how we can behave is determined by the abilities of our brain and other organs. Since the brain has an evolutionary history and has been formed over countless generations, it is obvious that human behavior is not only a result of postnatal learning processes, but also a result of the "phylogenetic history" of the human race. Evolution has set the framework for human behavior and for the behavior of other species such as cats, dogs, foxes, and birds as well.

The fact that human culture and human behavior are influenced by evolution does not mean we can find an evolutionary explanation for every behavior and for every cultural phenomenon. Evolutionary explanations for behavior and culture are often only speculation that we cannot verify. I find the trend that explains almost all human manifestations of life through the pursuit of selective advantage or the fear of Stone Age predators to be particularly problematic. Based on these explanations, the only reason for any interpersonal behavior would be for our own benefit. Altruism would then also be construed as a mere biological "behavior program," which serves the purpose of making us popular. Are we just nice to each other to guarantee help from other people or to make us more attractive to potential sexual partners? Cooperation would, therefore, be nothing more than the result of a "collective egoism" since cooperation entails advantages for each individual person. According to such hypotheses, people cooperate with each other for their tribe's or their species'—and thus their own—personal gain, but never for the sake of doing good. The idea that socially and ethically motivated deeds are nothing more than an evolutionary program is called social Darwinism.

One scientist who long sought for the sole cause of social action in evolutionary advantages is Edward O. Wilson, an internationally renowned professor of evolutionary biology at Harvard University. For a long time, Wilson, like other scientists, assumed that social behavior in humans is designed to increase the inclusive fitness of our species. That is, individuals behave socially so that the species *Homo sapiens* or their own community, preferably, survives. Are compassion and good deeds evolutionary mechanisms to prevent us from becoming extinct? Wilson has since changed his scientific opinion. His observations with animals taught him that social behavior could not be reduced to the mere preservation of inclusive fitness. "The old paradigm of social evolution, grown venerable after four decades, has thus failed," wrote the renowned evolutionary biologist. Inclusive fitness is "a phantom mathematical construction that cannot be fixed in any manner that conveys realistic biological meaning."[9]

Again, it is obvious that behavior has a biological and thus evolutionary *framework*. Social Darwinists, however, exaggerate with their speculations that can no longer be supported and tend to remind readers of fortune-telling rather than science. The most bizarre explanation for human behavior I have ever heard is still circulating among evolutionary psychologists. Are you one of those people who can't stand the sound of scratching chalk on the chalkboard? Many of us can't, and according to some psychologists, it is because the sound of chalk reminds us of macaques' warning calls in the jungle.[10] Our ancestors were supposedly made aware of dangerous predators by the cries of the monkeys. According to this presumption, we, therefore, perceive scratching chalk on boards as a shrill pain in the ears.

This explanation is obviously wrong for several reasons. First, the screams of the macaques do not sound like scratching chalk. These are merely personal associations of the psychologists who spread this theory. Second, the behavior of most people when they hear chalk scratching sounds does not coincide with the hypothesis that it was a warning signal for *Homo sapiens*. Most of us cover our ears and distort our faces. If we were on the alert, we would cock our ears and activate our perception organs instead of dampening them. I have never seen anyone who jumped up and ran away when they heard scratching chalk.

Does every detail of human feelings and behavior really have to be squeezed into an evolutionary context? Often such explanations say more about the imagination of the respective scientist than about facts. I am not claiming that we cannot find any causes for human behavior in evolution. However, our behavior cannot be explained solely through evolution, and we should use evolutionary explanations sparingly and deliberately. The resistance of sociologists to social Darwinism is justified.

Evolutionary medicine, as I have described it in this chapter, is above sociologists' criticism, as it does not look for explanations for behavior, empathy, love, free will, or culture in human evolutionary history, but rather explanations for illness and health. The evolutionary view of the human body is far less problematic and speculative than that of the human social life. Links between our ancestors' habitats and our body functions are easier to understand than our behavior and culture. Evolutionary medicine does not ask why we find music beautiful and why we do good deeds, but investigates, for example, the reasons cells turn into tumors or why modern life leads to such a dramatic increase in cardiovascular disease. Evolutionary medical observations of the human cell and the human body are far less associated with foolhardy conjectures than the evolutionary observation of social and cultural coexistence. Evolutionary medicine, therefore, offers much less fuel for public debate. Having said that, I will dare to interpret the interconnections between evolution, nature, and mental health in a later chapter of this book—sparingly and deliberately.

**THIS CHAPTER IN A NUTSHELL**

We are not only beings of culture, but also of nature. The emergence of many lifestyle diseases, which are dramatically increasing in the age of industrialization, can only be understood when we concern ourselves with the past of our species. Our species, *Homo sapiens*, has existed for 100,000 to 400,000 years, our genus *Homo* for a couple of million years. Most of the disease-causing viruses and bacteria emerged only after the transition from a culture of hunting and gathering to cattle-raising and farming. Since that time, more than half of today's infectious diseases were passed from animals to humans.

Our evolutionary past is most discernibly inscribed in our musculoskeletal system. Because of problems with my meniscus, I focused on a knee-friendly life instead of letting myself be coerced into a surgery. My pain disappeared through movement and exercise on a natural surface to which my human knee is adapted.

Studies have shown that many of the knee surgeries performed today are superfluous and even lead to long-term consequences such as joint inflammation. A Finnish research team published a clinical study in which placebo surgery proved to be as successful as actual meniscus surgery. Knee-friendly movement and building strong, healthy muscles that support the knee were the real factors for healing. There are, of course, cases in which an operation with partial removal of the meniscus is unavoidable.

In doctors' offices and clinics, knowledge of evolution has played a subordinate role so far. This is likely to change in the coming years. We know with certainty from studies that the drastic increase in nearsightedness is related to a lack of natural daylight during childhood. The eye needs its evolutionary, natural environment to develop. Many lifestyle diseases such as diabetes, high blood pressure, cardiovascular disease, metabolic issues, and others are related to the fact that our society has distanced itself too far from its natural habitat.

Evolutionary medicine also plays an important role in understanding the development of cancer and treatment of tumors. The risk of cells mutating is a part of life that occurred even in dinosaurs.

The primordial cell LUCA, the earliest ancestor of all creatures, was capable of reproduction by dividing itself. The only reason life could continue to develop and adapt to its habitat—that is, to evolve—is because genetic material is able to change. This ancient principle of evolution is still present in our body cells today.

Cancer begins with one cell that has genetic changes and that does not want to die. Environmental toxins increase the risk of this happening. Additionally, many people lack exposure to healthy substances of nature, which activate our immune system and protect against cancer. Therefore, tumors are an additional consequence of the broken relationship between humans and nature.

HIV is resistant to treatment because of rapid mutation, among other reasons. This is also evolution. The cure for AIDS might be possible in the future because science will understand the evolution of the virus better.

# 3

# HUMANS IN THE NETWORK OF LIFE

## A Part of the Forest

In the forest, you can get firsthand experience of how an ecosystem feels. The plant cover consists of several layers. Even the canopy itself is more than just one horizontal layer, but consists of at least three levels. The mighty crowns of the tallest trees form the top level. They are mostly the oldest and strongest ones that have made it to the top. They protrude out through the middle crown layer. Below, a third layer of trees can be found. It consists of those trees that didn't make it past the shade of the others. Many of them are waiting for a chance to push even higher when, for example, one of the tall trees falls and frees up space. The struggle for light is a natural selection mechanism that affects even the young trees. Only the nimble trees reaching for light come through. Of course, you cannot picture a layered structure of the forest as if it were drawn with a ruler. When viewed in cross-section, the individual tree crowns intersect with one another in a tessellated fashion. They fulfill important functions in the forest ecosystem.

The dense foliage keeps out the direct sunlight in the summer and creates a special forest climate. In the woods, it is a little cooler than out under the open sky and especially cooler than in the city, which is overheated unnaturally because of its bare asphalt and concrete surfaces. This is called the "urban heat island effect." Combined with muggy, polluted air, it is one of the reasons that midsummer in the city is more difficult to bear than in the countryside, especially for people with hypertension, breathing difficulties, and heart disease. Spending the summer in a mild forest climate is a real relief for those affected by these conditions.

Again thanks to the treetops, forest air also doesn't cool down as much as the air in open spaces during the night or on cold days because the heat is retained. Woodland climate is a very balanced climate. The forest floor does not even dry out during long periods of no rain, as it is in the shade and protected from evaporation by a layer of leaves, needles, and dead plant material. It provides a stable place for active soil life as well as an inexhaustible reservoir of nutrients and water for the trees and all other plants of the forest. The nutrients are in a constant cycle. The plants absorb them, then lose their leaves and needles, and in this way, restore the nutrients to the soil. Fungi, bacteria, and microorganisms of the soil take over the task of decomposing the plant parts as well as dead plants so that the nutrients are released.

In temperate climates, a couple inches of a dark humus layer forms in the forest and is rich in organic nutrients. This is the "storage room" of the soil. In tropical forests, the humus layer is less than half an inch thick because the trees and organisms in the soil are highly active throughout the year there. There is no winter break. The organisms in the soil decompose this layer, and the plants absorb the nutrients immediately. The food of the forest circulates in the tropics much faster than in moderate latitudes. It barely accumulates as a humus layer. For this reason, the rainforest areas cleared by the agricultural industry are not really suitable for farming. There are no nutrient reserves available. The nutrients are in the trees, and these are simply removed after clearing, to be sold somewhere on the other side of the globe as, for example, a luxurious piece of furniture.

The result of robbing the "lungs of the earth" is the necessity of then using massive amounts of fertilizers and pesticides so that the destroyed rainforest areas can be cultivated at all. In most cases, the crops are fodder for the meat production of industrialized countries. The fact that people do not behave as a part of the ecosystem does not change the fact that we—at least potentially—still have our place in it. I will return to this topic in a minute.

While a dense mulch layer doesn't accumulate in tropical rainforests because the nutrients are so quickly reused, in temperate zones, the winter creates the necessary vegetative pause so that material can accumulate on the forest floor. Everywhere on earth, foliage protects the forest interior from wind, which could blow away the soil and uppermost humus layer. Due to the shrubs and low plants on the edge of the forest, the ecosystem is protected from the wind all around. The crowns of the trees fulfill yet another purpose, which is important for the healing effects of all forests on earth. They keep the gaseous tree terpenes that strengthen our immune system in the forest air so that they do not escape from the forest. Because the trees provide shade, they also prevent the terpenes from being destroyed by the sun's UV rays on hot days. Measurements have shown that the concentration of terpenes in the forest is the strongest at three to six feet above ground, thanks to these protective mechanisms.[1] This is exactly the area in which our nose is located. This means we breathe air from the zone in the forest that is richest in tree terpenes.

"For me, this means that the trees want to be good to us," a woman told me who had read my book *The Biophilia Effect*. It was not the first time I had come across this interpretation. After explaining to one of my former book publishers that trees emit terpenes to which our immune system responds, he wrote a summary of our conversation. He wrote that trees emitted terpenes "when a person enters the forest."

When I talk with people about the healing powers of nature, I repeatedly hear the point of view that processes in nature are optimized just "for us." Some herbalists also assume that plants contain certain ingredients "for us" or that they form them to heal us. The idea that plants "come to us" when we need them is now a widespread interpretation of ecological processes in nature.

In the autumn of 2013, I visited Wolf-Dieter Storl and his family while researching food, gardens, and nature-loving people. A doctor of ethnobotany and a cultural anthropologist, Storl lives in the Allgäu Alps and is the author of numerous books on herbs and plants. Patches of fog accompanied us as we passed through the Storl family's garden in the morning, where I marveled at the countless varieties of fruits and vegetables. Along the way, we passed an area that was densely covered with teasels. These plants, known among botanists as *Dipsacus fullonum*, stretched their prickly stems toward the sky. At the tips were the already dried-up flower clusters.

"The teasels cured me of Lyme disease a few years ago," recalled Storl, pointing to the bushy plants. "Since it cured me," he went on, "it thrives here in my garden." From this observation, he drew a remarkable conclusion: "The spirit of the plant decided to stay with me after it had made me well." I thought it was the perfect opportunity to see the world for a few moments with different eyes that morning.

I once interviewed an herbal teacher—a trained expert on medicinal herbs—who told me about the insights she had gained during her work at a clinic. She said it is clear that in the surrounding meadows and forests, herbs always grow that are most needed at the clinic. The plants are therefore connected with humans somehow and "know" when we are in need of a certain active ingredient. They would respond by growing in the vicinity or multiplying very quickly.

Until his death in 2004, the herbalist monk Hermann-Josef Weidinger lived in Austria. He was known beyond the borders of his country as "Herbal Priest Weidinger." His successor is Benedikt Felsinger, alias "Herbal Priest Benedikt." During a walk through the mysterious forests around Geras Abbey in the region of Waldviertel, Austria, Father Benedikt explained to me that he too was convinced that the plant world responded to the current needs of the people in a region. Every year he observed how the most urgently needed medicinal plants could be found everywhere in nature. I have great respect for this interpretation of nature. It bears witness to a sense of interconnectedness with plants and reverence of life.

The fact that nature is downright soaked in healing influences such as tree terpenes and medicinal ingredients leads many naturalists to a certain anthropocentrism—that is, to the interpretation of the world predominantly from the perspective of human beings. The active ingredients of nature are present all the time and in nearly all ecosystems, but the mentioned plant experts and teachers pay special attention to those plants that are momentarily needed. Every year, nature offers its traditional herbal remedies against practically all symptoms. The only question is which of them we will focus on.

I had an experience similar to that of the mentioned herbalists with my own garden in the autumn of 2015. I was suffering from acute gastritis, which was extremely painful and unpleasant. The medication the doctor prescribed for me did not help. It was a proton-pump inhibitor, which temporarily impairs certain proteins that help pump stomach acid into the stomach. While taking this drug, the production of stomach acid is greatly reduced so that the inflammation can heal. But after two weeks, there was still no improvement. The treatment even made the symptoms worse because suppressing stomach acid has an impact on the ecosystem in the digestive tract. Too little stomach acid also means that the food in the stomach is not sufficiently broken down, and digestion is disturbed. Germs are killed less efficiently in the stomach because of the lack of acid, allowing them to attack the mucous membrane even more. They are also transported farther into the intestine, where they can upset the bacterial flora.

Proton-pump inhibitors represent a massive interference in the ecosystem of the stomach but are often prescribed for long-time use, especially in chronic gastric disorders such as reflux, when a damaged stomach valve leads to heartburn. The long-term side effects of these drugs are classified as dangerous by some doctors. Especially from the point of view of evolutionary medicine, which always sees humans and their functions in a larger context and looks for the cause instead of fighting symptoms, these kinds of drugs often do not represent a satisfactory solution.

Because the medication did not improve my stomach inflammation, I looked for an alternative. My search landed on the marigolds in my garden, which had returned unbidden after I had grown some the year before. I remembered my marigold harvest—the dried petals from the previous year were stored in sacks in my attic. I stopped taking the medication and instead drank marigold tea several times a day. I drank the tea slowly and lay down consistently after every second sip, alternating from the left side to the right, rolling on my back and my belly. That way, I could distribute the active ingredients over the entire stomach lining.

Marigolds have a very strong anti-inflammatory effect and are a perfect herb for the stomach. The long-proven active ingredients of marigolds are most concentrated in their yellow and orange flowers. These substances are flavonoids and—you guessed it—part of the terpenes group. My stomach inflammation improved noticeably within three days, and after a week it had largely subsided.

During this time, I suddenly noticed marigolds all around my garden. Everywhere around my house, the ground was full of them. They even grew in tiny crevices between the cobblestones directly outside my front door—they almost seemed to creep into my house. It can actually happen that a medicinal plant suddenly appears in a place where it hasn't been seen for years. The soil is full of different kinds of seeds that can lie dormant for decades before suddenly germinating and taking over the surrounding ground one day when the conditions are good and the competition is sleeping. Many plants also form offshoots, with which they can endure underground for many years, and then sprout unexpectedly. It is also common knowledge that birds, wind, and water bring new plant seeds as well. Plants have always conquered new habitats. Their sudden appearance can be understood in a botanical and ecological context.

Even if I prefer the ecological explanation and don't claim that plants "come to us when we are sick," I do not want to take away from the importance of the relationship between humans and plants. On the contrary, the numerous healing effects of nature on us—whether it is the effects of forest medicine or traditional herbal medicine—are another clear indication that we are creatures of nature. Tree terpenes and other plant substances are not new to our body. Our body knows what to do with them. We have evolved interacting with plants and as a part of the network of life. This is called coevolution. This kind of coevolution always leads to "mutual harmonization." From the perspective of coevolution, it is not surprising that our immune system reacts very positively to terpenes and that these terpenes are most concentrated in the forest at a height that is the best for our physique. Despite our distance from nature, we still have our place in the network of life. These ecosystems are our evolutionary home, which suggests—completely logically and without esoteric contortions—that nature has a healthy effect on us.

## All of Nature Pursuing the Forest

The layering of the forest goes far beyond the tree layers I mentioned earlier. In the shrub layer, we find bushes, small shrubs, and young trees, of which only a few will make it to the top. We also find an herb and a root layer in the forest. All of these layers are penetrated from below by climbing plants, which gives many forests a primordial appearance. The mosses and ferns of the forest are, evolutionarily speaking, the oldest rooted plants on the planet, dating back to the ancient forests of the Carboniferous Period, some 300 million years ago, and have changed little since then.

The largest living creatures that ever existed on earth are neither mammoths nor dinosaurs. They live in our forests, but we do not usually see them. The largest beings are fungi (mushrooms), which connect whole forest landscapes by means of extensive underground networks. They enter special symbioses with trees and interconnect with their roots. This way, trees and fungi exchange nutrients and water with each other. The tree provides the fungus with photosynthesis carbohydrates, while the fungus absorbs water and minerals with its wide-reaching network and shares these with the tree. This is a typical symbiosis. Mushrooms not only interconnect countless trees of the forest underground, but they also organize a highly complex transport of nutrients and water from tree to tree, shrub to shrub, and herb to herb.

The largest fungi interconnect almost 1,500 acres of forest, nearly 65 million square feet. Thanks to the huge mushroom network underground, the forest is connected into a single organism in which all plants and fungi are united. Biologists categorize mushrooms into the kingdom Fungi. The term "kingdom" aptly depicts the dimensions of the underground mushroom world.

Because the diversity and natural complexity within forests is particularly high, and we can literally immerse ourselves in forests, they are extremely attractive to us modern people. The woods are the counterpart to the asphalt surfaces of urban landscapes on which most of us move daily. Moreover, a forest is the essence of wild nature and represents a "desirable" condition for nature itself. The entire plant and animal world aims at forming forests all over the planet. In nature, there is a constant drive to create woodlands. If an ecosystem is left to itself, it reaches a stage of the highest complexity at some point, which then changes little for a very long time. Ecologists refer to this as the "climax stage." Where forests *can* grow, they *will* grow. If a forest cannot grow due to climatically unfavorable conditions, each ecosystem approaches a forest as closely as possible.

The climax stage is therefore not always a forest, but it is always "as much forest as possible." In high mountain ranges, above the forest line where the canopy is still dense, nature fills these areas with as many individual trees as possible. Even farther up, only small shrubs and trees such as juniper and mountain pines take root. This is called the "transition zone." The sparsely standing dwarf trees and dwarf shrubs of the Arctic and Antarctic tundra are the closest to a forest that is possible in such inhospitable climates. This is a battle zone in nature's pursuit of forestation. Savannas are lightly wooded with individual trees and groups of trees, between which an unbroken carpet of grasses and herbs grows. Even without the interference of humankind, there would be no closed forest cover in the savanna regions of the tropics and subtropics because there is a water shortage in these areas. The grasses and trees compete for this water and utilize all of the resources completely. They are in harmony with each other under natural conditions, and as many trees grow in a savanna as the conditions permit.

The pursuit of a forest state is noticeable, even if no closed forest cover is reached. Think of the taiga in northern continental areas. This biome of pines, larches, and spruces is an example of how nature can succeed in transforming even extremely inhospitable climate zones into the ideal of a forest. The taiga is also called "snow forest." In temperate lowland regions, extensive, stable ecosystems with broad-leaved and mixed forests would develop practically everywhere if humans suddenly disappeared. The higher the altitude, the more conifers would grow in the forests.

If nature is constantly striving to create forests, it is no wonder that we, as natural beings, also have a special place in our hearts for the forests of this earth. Nowhere other than in a forest can we so easily feel part of the network of life. It seems that almost every one of us has had a past experience with forests. I myself grew up on the edge of a forest, and my most beautiful childhood memories are tied to it.

When I called the University of Liverpool a few years ago, because I was thinking of starting a doctorate program there, I spoke to a very friendly secretary in the international study office. She asked me, out of curiosity, what I do for a living. "I write books about the relationship between humans and nature," I said. Immediately, she recalled a story from her life, which she told me with great enthusiasm. It was a forest story.

"I live just outside Liverpool," she said. "Every day after work I drive through a forested area. Ever since I started working at the University of Liverpool, not a day has gone by that I don't roll down the car window and slow down to let the forest air inside. I frequently pull over, too, and get out. That forest with its atmosphere helps me every day to leave work behind and really relax. It counteracts stress. I appreciate it very much."

## How the Earth Formed Our Organs

The Estonian biologist and philosopher Jakob Johann von Uexküll was one of the world's most influential researchers in zoology. Von Uexküll was born in 1864 and died in 1944. He distinguished between the *surroundings* and the *environment* of a living being. The surroundings are nothing but the space in which a living being or an object is located. An environment is much more than that. An environment is a space that is *influenced* by a living being, and at the same time, influences and shapes this living being.

There is, therefore, an interaction between the environment and living beings, through which the evolutionary process was made possible. In von Uexküll's words, every human being, every animal, and every plant is embedded in a "functional circle" with the environment. According to von Uexküll, we discern our habitat through our "perceptive organs." These are, for example, our eyes, ears, and nerves, as well as our skin with its ability to feel and perceive warmth or cold. In addition, our "operative organs" give us an opportunity to react to environmental stimuli and actively shape our habitat, for example, with our hands.

LUCA, the oldest common ancestor of all creatures, was interwoven with the environment through a functional circle. LUCA's abilities to interact with its environment were, of course, not as advanced as those of today's humans, animals, or plants. However, the primeval cell was probably capable of both influencing its environment by secreting simple chemical substances and reacting to environmental influences with movements, ingestion of nutrients, and so on. It had simple perceptive and operative organs.

Biologists call microscopically small organs of present-day unicellular organisms "organelles." Many single-cell organisms have an eyespot, with which they perceive light and darkness. There are also organelles that serve for digestion. Unicellular organisms can even ingest and digest food and then eliminate their metabolic products. They just latch on to a nutritious particle in the water and push in their outer perimeter to incorporate it. This is a very simple form of "swallowing." In the cell, the food particles enter a separate cavity with digestive enzymes. This principle is the same as what happens in our stomach and intestines.

Movement organelles, which look like hairs or whips under the microscope, can move single-cell organisms in the water to specific points. They are attracted by light or nutrients and emit messenger substances to find each other and to fuse together during "unicellular sex." Unicellular life-forms are by no means simply minute dots in the water that passively float back and forth. In their microscopic water world, they interact actively with other single-celled organisms and their environment.

During the course of evolution, more and more specific organs were formed with which living beings could enter into a functional circle with the environment. The habitat of our ancestors shaped our present organs. We should be understood as complementary to our natural habitat, that is to say, as a piece of the puzzle that fits with the others.

The bottom line is that our ancestors did not adapt passively to their environment. Evolution theory has so far assumed that the adaptation is due to *accidental* damage to the genome alone. The DNA damage leads to mutations—that is, changes in the genetic material of a cell. Many mutations have no consequences for the living being; these are the neutral mutations. Most mutations do not benefit the affected organism, which is to be expected in the case of accidental damage to the DNA. In many cases, mutations even lead to a disadvantage for the living creature or its offspring. These are the negative mutations. The inherited color blindness of humans is, for example, the result of a mutation during the course of human evolution. Sickle cell anemia, in which the red blood cells absorb less oxygen because of their altered form, also comes from a negative mutation. Some negative mutations even lead to death.

A very few mutations are positive and lead to an advantage. For example, there is a small number of people worldwide who are resistant to HIV. Behind this is probably a mutation. Medical geneticists are working to find out which mutation makes a person resistant to HIV. They hope to find a new treatment with this discovery.

For our considerations on the evolution of species, it is relevant that some positive mutations have led to adaptations to the environment. The fin is an adaptation to the habitat of water, and the wing an adaptation to the world of the birds. Living creatures with positive mutations were better off in their habitat than the rest of their species and therefore prevailed. Nature selected them over the less adapted ones.

Can the "inventions" of nature, such as fins and wings, really be explained by mutations that were purely coincidental? I remember the first semester of my biology studies in 2000. Our zoology professor thought it was extremely unlikely that accidental changes in DNA could lead to a mutation that would be a useful environmental adaptation for the living being. He compared the evolutionary process with a computer simulation that works with virtual DNA sequences. According to our professor, a current computer (at the time) with a random number generator would have to calculate for a few million years, simulating random DNA damage, until one of these coincidences led to a positive environmental adaptation.

After his calculations, the professor did not come to the conclusion, however, that maybe a few pieces of the puzzle might be missing to understand evolution. His argument was rather: "It is extremely unlikely, and yet that's how it happened." He represented the mainstream view that life on earth had had enough time (more than four billion years) to evolve based on an accumulation of *accidental* DNA damage.

Feelings of resistance stirred in me. I entered a discussion with the disputatious biologist. If the calculation of a single positive mutation had already overwhelmed a random number generator from the year 2000, how could it be the explanation for such a rich biodiversity that we find on earth today? After all, not only one, but an unbelievable number of positive adaptations were necessary in order for our ancestor, LUCA, to evolve into one hundred million different forms of life. The transition from a unicellular to a multicellular organism and from aquatic creatures to land dwellers, the return of mammals to the sea through an almost unimaginable transformation of wolf-like predators to whales and, finally, the emergence of humans trying to reconstruct their own natural history—all of these are achievements of adaptation that require an unimaginable cascade of single positive mutations, not to mention the fact that they had to happen in a certain order and not just randomly.

The idea that the turbulent history of life on earth and the entire biodiversity of the world are ultimately due to mere accidental damage to DNA or to copying errors in cell division already seemed unbelievable to me in 2000. My zoology professor pointed out to me at the time that my skepticism sounded similar to the position of the creationists or at least gave them more ammunition. They notoriously question the theory of evolution and instead believe that each individual species was created by an intelligent divine designer. They believe that small changes within a species are possible, but not the emergence of new species through natural processes; only a creator is able to create a new species, they argue. Human beings, trees, wolves, elephants, fish, and roses were all "made" by the intelligent designer, much in the same way it is written in the Bible.

That explanation for life processes, however, was far from what I meant. I did not want to express anything religious. In my naive brain as a university freshman, it was not easy to imagine how in the world arbitrary damage to the heredity material of LUCA was supposed to have led me to sit in a lecture hall four billion years later, arguing with my professor.

At this point, my teacher fell back on the tactic that biologists always use when a phenomenon in the history of life is so unlikely that we actually have no explanation for why it happened. This tactic is presenting the "anthropic principle," which states that it does not matter how improbable a certain phenomenon is that occurs in the course of evolution if we need to base the formation of species on it. In other words, it must have happened so; otherwise, we humans would not be here to think about this phenomenon. In many cases, the anthropic principle is basically nothing more than an elegant way out if scientists do not want to admit that they do not know *why* something was as it must have been according to current knowledge, or if they don't know *how* it could have happened.

My professor explained to me, according to this principle, that it was indeed extremely unlikely that accidental damages to the genetic materials could have led to the formation of the two of us, but it must have been so because otherwise, we would not be here to think about it. After all, we—you and he and I—came into being because the improbable occurred. What I could not explain to my debate partner at the time, who for me was an authority on the subject, was that I had no doubt that biodiversity originated through the evolutionary process. I also did not doubt that the shaping forces of evolution are based on mutation and selection. I merely had trouble believing the then-accepted view that successful environmental adaptation was due solely to accidental damage to the genetic material or to errors in the tiny cellular copying machine. This would mean that living beings would be in a completely passive process of formation and change and would *depend* on the chance occurrence of DNA damage in order to adapt to their environment. This is exactly what many evolutionary biologists believed up until a few years ago, and some still believe it today.

Jakob Johann von Uexküll was of the opinion that living beings bring adaptations to their environment upon themselves through their behavior and their interaction with the environment. Accordingly, environmental adaptation is also an *active* process of the organism, or the organism at least partially influences the adaptive process. Evolution thus becomes an act of living creatures—that is to say, a manifestation of life. In other words, we are not "subject to evolutionary processes," as it is often expressed in biology, but we are *involved* in the evolutionary process. That would mean that our own behavior toward the environment is passed on to subsequent generations and influences the evolution of our species.

The assumption that environmental adaptation and evolution have something to do with an active influence of living beings is supported by the latest findings of epigenetics and is based on scientific evidence. Geneticists and developmental biologists have found, in humans and animals, some features in appearance, behavior, and bodily functions that cannot be explained by conventional genetics. The evidence is mounting that experiences and environmental influences are passed on to offspring through mechanisms that are not yet understood—that they can manifest themselves as "imprints" in our genetic material. This contradicts classical genetics, in which scientists assumed until now that any genetic material of a living being is unalterable after birth. Geneticists were convinced that the experience of a human being, an animal, or a plant could not affect its offspring through heredity, but rather only through psychological and social mechanisms at most during upbringing.

However, several medical studies have raised doubts about this position. The geneticist Marcus Pembrey from London and the medical professor Lars Olov Bygren at the Umeå University in Sweden found in a large-scale statistical study of men in the Swedish town of Överkalix that the risk of developing diabetes was linked to the nutritional situation of the men's grandparents during the grandparents' *childhood*.[2] The study was published in the *European Journal of Human Genetics* and in the renowned science journal *Nature* and shows that the environmental conditions under which humans live can affect future generations. Because this phenomenon cannot yet be explained by the established laws of genetics, it is referred to as "epigenetics," that is, as "additional" to the known processes of genetics. Epigenetics has revealed mechanisms by which certain genes are activated or deactivated in the cells of living creatures. A chemical marker connects to the gene and switches it on or off. Evidently, environmental experiences of a living being can directly affect the genetic material.

A few decades ago, biologists found epigenetic phenomena for the first time in honey bees. All the worker bees of a colony are descendants of the queen. Under normal conditions, the larvae that the queen lays become workers. But when a colony begins to swarm and divide, one of the two colonies is without a queen. A queen can simply die sometimes, too. The bees breed a new queen whenever she is missing. They take a conventional larva, which they feed with royal jelly. This is a special gland secretion of the bees. In addition, worker bees place their chosen royal larva in a vertical cell of their hive, and not, as usual, in a horizontal one. They therefore create special environmental conditions for the larva through the feeding and the housing, which leads to the hatching of a new queen. It is the exact same genetic code in the larva, but depending on the environmental influences, it either becomes a worker bee or a queen.

The environment, therefore, decides which genes are activated or deactivated. A queen looks different from the workers and behaves differently, even though there is no such thing as a queen's DNA or worker's DNA. The genetic code seems to be *interpreted* to a certain extent by the living creatures. Some biologists compare this process with the notes in music. The notes are written down, but the interpretation sounds different with every musician. In this way, it is possible to explain why monozygotic twins, who are genetically identical, do not remain clones but develop into clearly distinguishable individuals and characters, and who over the course of life often differ physically from their twin as well.

Marcus Pembrey conducted several studies on epigenetic inheritance in humans and, together with other scientists, came to the conclusion that our environment can have a direct impact on the activity of our genes. In an essay published in *European Journal of Human Genetics* and in *Nature*, Pembrey wrote about the processes that he assumes occur with the woman's egg cells and the man's sperm cells. Information from the environment might be imprinted on our sex cells at their formation and later passed down to the offspring during conception.[3] Pembrey deduces this from his investigations and experiments.

Jonathan Seckl, a professor of molecular medicine at the University of Edinburgh, also considers the findings of epigenetics to be important. During an interview with the BBC, Seckl said, "These results are provocative. Some find them difficult to accept. But it's quite clear now that a number of laboratories are finding similar findings in the various systems that they are interested in. So the phenomena are there."[4]

The exact mechanisms by which environmental experiences affect epigenetic inheritance are largely unknown at this time and will likely be a mystery for scientists for a while. The use of epigenetic inheritance for evolution is, however, clear: epigenetic imprinting helps the adaptation from one generation to the next. It makes living beings active participants in the future of their species. Perhaps epigenetics is one of the missing pieces to the puzzle that explains how a hundred million species on earth have been able to adapt so efficiently to their habitats over four billion years since LUCA. After reading these findings, my former zoology professor would probably select another comparison instead of a random number generator as the driving force of evolution. Random damage and copying errors in the DNA are not sufficient explanations for highly specialized environmental adaptations. Epigenetics fills a great gap in the understanding of biology, and we are just getting started.

In the same BBC documentary, Pembrey emphasizes that his findings point to the possibility of actively participating in evolutionary processes. "The first time one had a photograph of the Earth—this delicate thing sailing through the universe—I'm sure it had a huge effect on that sort of *save the planet* feeling. I'm sure that's part of why the future generation think in a planetary way—because they have actually seen that picture. And this might be the same. It may get to a point where they realize that you live your life as a sort of guardian of your genome. You've got to be careful with it because it's not just you. You're also looking after it for your children and grandchildren."[5]

## Is Earth Itself a Living Organism?

The findings of epigenetics could help revive Jakob Johann von Uexküll's idea of the functional circle, in which living beings interact actively with the environment through their operative and perceptive organs. Von Uexküll also said that objects have borders, but not animals, plants, or humans. He refused to establish the boundaries of living creatures at their physical external borders. An animal does not end at its skin surface or at its fur, nor a tree at its bark.

"Are the dandelions in the meadow one plant, joined as they are through a common root system, or many?" Alva Noë, professor of philosophy at the University of California, Berkeley, asked this question in our own century. In his book *Out of Our Heads*, he asked similar questions: "Is Macintosh OS one program, the operating system, or is it many? . . . Is Williams-Sonoma one company or several?"[6] Noë concluded that where we draw the borders is an arbitrary decision.

In the network of life, these questions are even more difficult to answer than with companies like Williams-Sonoma. Is the forest, in which all forms of life are connected by a gigantic subterranean fungus, a collection of many organisms, or does it represent a *single* organism? After all, the life-forms of the forest are not simply interlinked. They create a functional network in which nutrients, water, and communication molecules are exchanged. Our organs also do this with each other. Recently, biologists have observed that plants can even emit subterranean "clicking" sounds that spread as electromagnetic signals but are not audible to our ears. This is most likely another internal communication channel of the forest.

The forest behaves like an organism. Similarly, the human body is formed through the connection of our organs and body cells. Couldn't we apply this line of thought to humans so that we see our bodies as a collection of many microscopically small life-forms that, together, yield us? Nature's functional circles combine everything with everything else and everyone with everyone else.

Some natural philosophers and biologists put forth the argument that the earth itself is a living organism. This is known as the "Gaia hypothesis." Gaia was earth personified in ancient Greek mythology. The Gaia hypothesis came from American cell biologist Lynn Margulis at the University of Massachusetts and British biophysicist and physician James Lovelock at the University of Oxford. The hypothesis later emerged in a distorted form in popular esoteric books in which Margulis and Lovelock's theories were used to spread the notion of a sensitive earth that not only "consciously acts," but also "suffers" as humans and animals can. Some esoterics took the Gaia hypothesis to mean that the earth was a kind of deity, which according to some portrayals, even intervenes, rewarding and punishing during happenings on the planet.

These distortions scientifically discredited the originally respectable Gaia hypothesis. Lovelock stated explicitly that he was not referring to a sensitive planet with consciousness and intentions, nor an earth as a goddess. "When I speak of a living planet, it should not have an animistic connotation. I do not think of a sentient earth or stones that move according to their own will and purpose." Lovelock emphasized that the Gaia hypothesis has a scientific perspective. He also made it clear, though, that he did not condemn those people who long for an earth with a soul. "I respect the attitude of those who find solace in the church and speak their prayers, while at the same time they concede that logic alone does not provide a convincing reason for believing in God. In the same way, I respect those who find consolation in nature and may wish to speak their prayers to Gaia."[7]

The earth is a holistic system that organizes itself. Let's consider ecosystems in a larger context. A forest ecosystem does not end at the edge of the forest, but there is a lively exchange of material between the interior of the forest and the surrounding areas. The composition of soil life and soil chemistry does not change abruptly between forest and meadow, but rather changes gradually. The dark humus of the forest gradually changes into the lighter brown humus of the meadows. The forest climate influences the climate of the surrounding country-side and vice versa. A forest ecosystem is therefore embedded in a network of ecosystems between which there are no clear dividing lines. Forest edges are transition zones, not boundaries.

In nature, everything is constantly interacting. The fact that regional climatic changes always affect the climate as a whole makes it clear that the earth is a single, large ecosystem—the "geo-ecosystem." Within the framework of global warming, the temperature can rise in certain areas, but in others, it turns cooler as a result, for example.

The Gaia hypothesis states that the earth with its biosphere—that is, the totality of all living creatures, including us humans—can be regarded as a living organism. According to the hypothesis, this living organism maintains conditions for long periods of time that make life in the biosphere possible along with the development of this life—that is, evolution. This is the key point of the Gaia hypothesis. The conditions under which life can arise and develop are limited. On earth, these conditions have been preserved by complex equilibria for eons.

The earth is also capable of reacting dynamically to the influences of the species that it hosts. Seen from space, the earth seems comparable to a living being in constant dynamic motion, especially in time-lapse photography. In earth's cycles, Margulis and Lovelock see physiological processes. This is what biologists refer to as life manifestations that ensure the functioning of an organism. Our digestion and the transport of substances in our blood vessels are examples of physiological processes. According to the Gaia hypothesis, rivers transport substances in a similar way as our bloodstream. Rivers are therefore the "veins" of the organism Earth. The individual ecosystems—for example, a "breathing forest"—are like organs. A forest would, therefore, be a kind of "lung," and tree crowns, with their branches, twigs, leaves, and needles, indeed evoke an image of the bronchi.

## Humans Don't End at the Surface of Their Skin

For our considerations, it doesn't matter whether we share those conclusions of the Gaia hypothesis and look at the earth as if it were a living organism. The observations on which the hypothesis is based are of importance with or without "Gaia." In nature, everything is so closely interconnected that it is impossible to draw clear borders. And this applies to us humans as well.

We are in a biochemical exchange with the environment. This also supports Jakob Johann von Uexküll's functional circle of nature. The tree terpenes from the forest air and their effects on our natural killer cells are examples of how influences from other life-forms integrate into the processes of our own body through the functional circle. Tree terpenes become part of the functioning of our immune system when we inhale them or come into contact with them through our skin and mucous membranes. Microorganisms that colonize our intestines make a healthy functioning of our digestive system possible. It is probably not news to anyone that the human body is a habitat for bacterial populations, which in turn maintain our own life functions. Relatively new, however, is the discovery that microbes in the intestine can biologically communicate with our immune system and our brain. We are ourselves a kind of "ecosystem." Just like the forest, we cannot draw a clear line between ourselves and our environment. We are integrated into a functional circle with nature. Moreover, we set the limits of our organism, ourselves, by extending or limiting the section of our environment with which we interact.

In 1998, neuroscientists Matthew Botvinick and Jonathan Cohen of Princeton University published in the journal *Nature* the results of their famous experiment under the title "Rubber Hands 'Feel' Touch That Eyes See."[8] The experiment was later repeated several times by other scientists. The test subjects sat at a table and laid a hand on their lap so that the hand was hidden by the table. A rubber hand lay on the table. There was no connection between the rubber hand and the participants. A staff member now began to brush the rubber hand with a paintbrush so that the test subject could watch. Another researcher brushed at exactly the same time and in the same way with a second brush over the subject's real hand, which was hidden under the table. After a while, the subjects felt that they were experiencing the touch in the rubber hand on the table. They perceived this rubber hand, in spite of the knowledge that it was not part of their body, as their own hand.

In a second stage of the experiment, the subjects also hid their second hand under the table. The touches of the rubber hand continued in sync with the touches of the first hidden hand. With the second hand, the test subjects were asked to point in the direction of the touch they were feeling. They all pointed to the table, at about the place where the rubber hand lay.

Alva Noë came to an interesting conclusion about the rubber hand experiment. "The rubber hand is, of course, not part of us. But that is not because it is made of rubber or not connected to our body. Crucial is that the rubber hand has a different destination than us. Our survival depends on the rubber hand only superficially and coincidentally. Our own hand, in contrast, is reliably integrated into our sensory and motor interactions with our environment and our other sensory experiences." This formulation of the philosopher suggests that objects that are not part of our body can become part of our organism under certain circumstances. "If it were possible," Noë continued, "to integrate the rubber hand into an active interplay between the world and the body, the rubber hand would be a part of us in this sense."[9] This assertion has been confirmed several times in practice.

A British-born Irish and Spanish artist named Neil Harbisson became the first cyborg in the world.[10] This is what we call a mixture of human and machine. Harbisson has been color-blind since birth. At the age of twenty, he had a color sensor attached to his head that is directed forward as a third eye. A computer chip converts colors into audible tone frequencies, which are transmitted to his ears via the skull plate. After he got used to it, Harbisson learned to interpret these quiet tones. Since then, he has "heard" colors. His body has integrated the cyborg eye into the body schema and made it a functional perceptive organ for environmental stimuli—to say it in Jakob von Uexküll's terms.

The physician Carter Collins, together with his team from the Presbyterian Medical Center in San Francisco, helped blind people in a scientific study to "see" without eyesight. A small camera converted pictures into a kind of palpable image. For this purpose, the researchers attached a square plate with small metal pins onto the back of the test subjects, which they felt on their skin. The images from the camera were represented in real time by the metal pins. We can imagine this to be something like images on a small computer screen made of pixels. Each pin represented a pixel. In this way, the camera transmitted contours and light and dark shadowing. The darker the respective pixel, the stronger the pressure of the metal pin corresponding to this image position. After a period of adaptation, the participants' skin became sensitized, and they were able to perceive images in their mind by means of the pattern of the metal pins they felt. They even managed to recognize obstacles and move around a room without colliding with objects. They learned in a way to see with their skin. They had developed a new perception organ and integrated it into their body.[11]

I, like everyone, have experienced similar phenomena in everyday life. When mountain biking on forest roads, for example, I can feel the texture of the road surface through the wheels. The more I rode my mountain bike, the more I could use it like a body part. I became more and more sensitive to contact that actually took place under the tires, and I could perceive more and more details of the ground. I had integrated the bike to some extent into my body schema—at least as a temporary perceptive organ. Ever since this integration, it has been as if I can "feel" with the rubber of the tires.

When we touch a rock with a walking stick, we feel the rock at the tip of the stick. Every driver knows this phenomenon as well. Over time, we develop a physical feeling for our car. We feel the outer boundaries of the car while driving as if they were extended boundaries of our own body. We know exactly where the rear bumper ends, even though we do not see it. We "feel" it. According to Jakob von

Uexküll, bicycles, cars, skateboards, and other vehicles become perceptive organs of our body after regular use and simultaneously become operational organs, with which we influence our environment and move through it using them.

Every organism—whether human, animal, plant, fungus, or single cell—is in a permanent exchange with its environment. The life process constantly transcends the physical surface and forms an overarching system with the environment. Our body is therefore very flexible with its external borders. Our body schema is not static and extends far beyond our skin. This is important for understanding how we are interconnected with nature, as is the fact that there is a biochemical exchange between us and plants. For epigenetic inheritance and the active adaptation to the environment during the course of evolution, this functional circle of human beings and environment is essential.

## Biophilia and the Healing Code of Nature

In the 1960s, the great German-American psychoanalyst Erich Fromm presented his ideas about the opposing forces that are deeply at work in the human psyche in his famous work *The Heart of Man*. With his ideas, Fromm moved close to the drive theory of the Viennese Sigmund Freud, the founder of psychoanalysis. Human existence, as Freud put it, represents a constant struggle between Thanatos, the death instinct, and Eros, the life instinct. In a healthy human being, Eros retains the advantage in this game of forces. In a similar way, Erich Fromm analyzed the hidden processes in the human psyche. In the case of Fromm, it is not Thanatos and Eros that confront each other, but necrophilia and biophilia.

Derived from ancient Greek, "necrophilia" means "love of the dead," and "biophilia" means "love of life." Fromm wrote that every human being carries necrophilic forces inside them that are responsible, among other things, for acts of violence and illness. In a manner similar to Freud, he assumed that biophilia maintains the upper hand in a healthy human being. "The tendency to preserve life and to fight against death is the most elementary form of the biophilous orientation, and is common to all living substance."[12] That means biophilia must also have existed in unicellular organisms and even in LUCA as a principle of life and a will to live.

Fromm also spoke of human biophilia, which goes far beyond the elementary will to live. "The person who fully loves life is attracted by the process of life and growth in all spheres."[13] These life and growth processes are everywhere in nature. We only need to think of a forest or garden. The biophilic forces in human beings thus manifest themselves through our *attraction*, as Fromm writes, to nature and its forms of life. I would like to leave no doubt that this form of biophilia is also at work in the psyche of animals; I am sure of it. In a way that is difficult or even impossible for us to discover, it might even be at work in plants. Through biophilia, humans, animals, and plants take part in the whole of nature. Through processes deep in our psyche, biophilia connects us with the network of life from which we come.

Fromm's idea of biophilia also attracted the attention of natural scientists. One of them was Edward O. Wilson. Wilson was the first to introduce the idea into biological sciences. If all, or at least most, people have an innate love of living things in nature, this characteristic must be related to our evolutionary history. Wilson published his "biophilia hypothesis" in the early 1980s, shortly after Fromm's death. It supposes that our attraction to nature is genetically predetermined. According to this hypothesis, people have an "urge to affiliate with other forms of life," as Wilson put it.[14] Biophilia is the human bond with other species.

The term "biophilia hypothesis" is somewhat misleading. The reason biologists speak of this as a hypothesis (that is, an assumption) is because it cannot be scientifically proven whether biophilia is really an innate force that is genetically predetermined in humans. Serious evolutionary biologists are generally cautious with their assertions since they cannot travel millions of years into the past to examine their ideas about the course of evolution. It is equally impossible to expect that paleobiologists will one day excavate a piece of petrified "love of nature" from a cave dweller to prove that biophilia is the result of evolution.

After all the observations we have made so far regarding the origin of humankind, however, it would be rather surprising if we did *not* have biophilia. If the environment of our ancestors is firmly written in our biological body, then it must be in our psyche as well. Our evolutionary past with nature is "submerged in our psyche" and now works "from the depths," to quote phrases from psychoanalysis. It affects us from our unconscious and even influences our dreams and our aesthetic perceptions.

Gordon Orians, professor emeritus of biology at the University of Washington, Seattle, is a globally recognized tree expert. He owns a large photo archive with pictures of trees from all over the world. Orians used these photographs to examine which tree forms people intuitively prefer—that is, which trees we find most beautiful when we have to decide spontaneously. He noted that people unconsciously fall back on three rules. We like trees that have branches good for climbing better than those that give us little support. We also like trees with crowns that give us shade. Trees with expansive crowns are therefore particularly popular in parks from an aesthetic point of view. And lastly, we are attracted to trees that bear edible fruits or contain substances that keep us healthy.[15] Thus, it appears that the evolutionary past is reflected in our aesthetic perceptions. We intuitively recognize which trees have always been good for us humans.

No matter where we live, we modern people carry a piece of savanna in us even today in the twenty-first century. The transition from *Homo erectus* to *Homo sapiens* at least 100,000 years ago, and probably even 400,000 years ago, took place in the savanna landscape of East Africa. Biologist John Falk, a professor at the University of Oregon, surveyed residents of the Northeast about which landscapes they preferred. Do we have a preference for landscapes where we grew up as children? Or are we born with an intuitive preference for a specific kind of landscape? John Falk showed the participants hundreds of pictures from different vegetation regions on earth, and the participants were supposed to decide spontaneously where they would feel most comfortable.

The savanna was a clear favorite for the children. Older inhabitants found the savanna at least as appealing as the landscape they were accustomed to, while many liked the savanna even better. Professor Falk conducted the same study in 2009 with residents of a rainforest in Nigeria. The results were surprising: even the rainforest inhabitants intuitively preferred the savanna landscapes to the rainforest.[16] Our evolutionary history must play as much a role in our aesthetic sense of nature as does the region in which we grew up. The memory of savannas is imprinted deep within our unconscious and has shaped our biophilia. Of course, we also have to bear in mind that survival in the jungle is associated with hardships and dangers we are not familiar with in forests of temperate latitudes. This could be one reason that even rainforest inhabitants respond so positively to savannas.

The fictional inhabitants of the earthlike moon Pandora from the movie *Avatar* look back on a coevolution with blue plants. If these beings really existed, they would certainly have biophilia. Similar to how we are attracted to green trees, they would react to blue ones. The biophilia of marine life—and I am sure that fish also carry biophilic forces within—is certainly not gauged toward savanna trees, but toward the underwater world of life in which the evolution of the fish took place. A fish would not know what to think about a tree. Its species has no concept of trees.

Psychologists who focus on creativity have established beyond a doubt what most people know from experience: being in nature encourages creativity. Simple tests show that we are much more creative in nature than in enclosed spaces. One person who experienced it firsthand was Michael Jackson. In an interview with the British TV channel ITV2, he stated that many of his biggest hits were written in the crown of a mighty old tree on his property: "I've written so many songs from this tree. I wrote 'Heal the World' in this tree, 'Will You Be There,' 'Black or White,' 'Childhood.'"[17] There was a sparkle in The King of Pop's eye when he said it. He called the old tree his "giving tree."

When we are in nature, biophilia becomes active and spurs our mental processes. The biophilic forces within us drive us to be close to life and growth processes. This quest is satisfied in nature. It is not surprising that our biophilia stimulates our creativity, our expression of life. But biophilia can do so much more.

Being in nature is not only inspiring, it also has medical and psychotherapeutic potential. By experiencing nature, we place our body in the original functional circle made of humans and the environment from which we emerged. We put the two matching puzzle pieces together—us and nature—into one whole. Nature-human medicine can keep us healthy and support us in the fight against diseases. These effects can be described strictly scientifically as evolution-based, health-promoting effects of nature experience on humans. I refer to this simply as the "healing code of nature." Our biophilic forces respond to this natural code, which they intuitively detect in the environment and decode. This interplay is itself the result of our functional circle with nature for millions of years.

The healing code of nature will be the focus of the next chapters, where I'll be talking about its concrete medical and psychotherapeutic applications.

## THIS CHAPTER IN A NUTSHELL

Trees protect their own terpenes from direct sunlight in the forest and prevent these substances that are healthy for us from being destroyed by UV radiation. By retaining their own gaseous plant substances so that they cannot escape, these compounds remain concentrated at precisely the height at which our nose is located. Are the trees trying to be good to us? The fact that nature offers healing substances for almost every disease causes some of us to feel that natural remedies have been made by plants "for us." This is an anthropocentric point of view. However, every ecosystem is permeated by active ingredients, most of which we don't even know about yet. Because of coevolution between humans and plants, these substances are not new to our body and have a direct effect on it. This is how we can recognize that we have a permanent place in the ecosystems of earth.

In biology, evidence suggests that living beings actively participate in the evolutionary process. Through their interaction with the environment, they influence further development of their organs as well as their entire species. Epigenetics supports this assumption with its latest findings. Our environmental experiences can have a direct impact on our genetic code within our lives by, for example, switching individual genes off and on.

Recent findings from biology and neuroscience show that the human body does not end at its skin surface—and the same goes for animals and plants. Besides the exchange of substances between us and the plant world, we are also noticing that our body is very flexible with its external boundaries. Our life processes constantly exceed our physical body. We are even able to incorporate foreign objects into our body schema—like a walking stick or a bicycle.

We are, after all, a kind of ecosystem, an accumulation of body cells and organ systems, with many microorganisms that colonize our digestive tract and our skin. Just as the forest has more of a transitional zone instead of a border at the edge, we cannot draw borders along our natural habitat either. The intertwining and nesting of ecosystems can be traced so far that the whole earth can be viewed as one single ecosystem, of which we are a part.

We are integrated into a functional circle with nature. A force developed in humans from this integration, which the psychoanalyst Erich Fromm described as "biophilia." This is the love of life and growth processes. The evolutionary biologist Edward O. Wilson was the first to introduce Fromm's concept into the natural sciences. His biophilia hypothesis states that our attraction to nature is genetically predetermined in us—that is, it is innate.

Nature is full of influences and stimuli that are healthy for us and that we can decode thanks to our biophilic forces. That is why I refer to the "healing code of nature." It is itself a product of our evolutionary history.

# 4

# ECO-PSYCHOSOMATICS

## From Anticancer Terpenes to Natural Heart Protection

Qing Li, the medical researcher I mentioned earlier, conducted studies with a large team of scientists on the effects of a forest on the human heart. The researchers divided their subjects into two groups. One group went for a walk through the forest while the other walked in the city. Both went the same distance and at the same pace. The researchers made sure that the type and intensity of the movement was the same in both groups. The only variable was the environment: city or forest. Before and after the walk, the scientists took a blood sample from the participants. The walk in the forest led to a significant increase in a substance called dehydroepiandrosterone (DHEA) circulating in the bloodstream. This effect did not occur in the urban walkers.[1]

DHEA is a steroid hormone that is formed in humans and many vertebrates in the adrenal cortex. DHEA is actually the most common hormone produced there. It is produced in a part of the adrenal cortex that biologists call "zona reticularis," a net-like zone. DHEA leads to both the female sex hormone estrogen as well as the male sex hormone androgen. It can behave like one or the other. DHEA lowers the energy consumption of cells, which is why some scientists believe that it has a life-prolonging effect, but that hypothesis is not proven. On the other hand, the effect of DHEA on our blood vessels and the heart has been verified multiple times.

The DHEA content in our blood reaches its peak around our midtwenties and then decreases continuously. This decrease is likely to be one of the many, little-researched biological reasons why we age, and the risk of illness increases with age. At age seventy, the DHEA content in our blood drops on average to 20 percent of its peak.

When we take a look at the many effects of this steroid hormone, it becomes clear why its decline as we age is associated with an increased risk of disease. DHEA is involved in the regulation of the blood glucose levels and maintains cardiovascular functions. And how does it do all that? When compared to a placebo in a clinical study, the administration of DHEA significantly improved the function of the epithelial cells of the blood vessels. The epithelial cells line the insides of blood vessels. They form several layers and play an important role in contracting and relaxing the blood vessels. Doctors have long known that the risk of heart and circulatory diseases increases when the blood vessels do not relax after contraction—that is, they "harden." The resulting constrictions hinder the flow of blood through the body. Clinical studies have shown that the administration of DHEA to patients with cardiovascular disease and at risk of cardiac problems led to clearly measurable improvements in the permeability of blood vessels after four weeks. After twelve weeks, the elasticity of the vessels doubled.[2]

Numerous other studies have been conducted that demonstrate the effect of DHEA on the heart and circulation.[3] DHEA reduces the risk of developing blood clots (thrombosis). It counteracts the so-called deadly quartet, which is what physicians call four major findings associated with the risk of dying of a heart attack: high blood pressure, high level of triglycerides in the blood, increased blood glucose levels, and obesity. Above all, DHEA counteracts coronary heart disease. When plaque builds up in the coronary arteries (arteriosclerosis), these arteries lose their elasticity and can restrict blood flow. The heart muscle is no longer supplied with an adequate amount of oxygen.

Coronary heart disease is one of the most common causes of premature death in industrialized nations.

At this point, I would like to present a brief overview: The heart-protecting substance DHEA from the adrenal cortex increases during walks in the forest, but not during walks through the city. The hormone counteracts cardiovascular diseases, which occur especially in industrial societies and are called "lifestyle diseases." So, once again, the questions arise that I have already asked in a previous chapter: Could it be that not only unhealthy industrial food, lack of movement, and harmful environmental influences play a role in the development of these diseases? Could it be that removing ourselves from nature also contributes to these diseases because we are no longer in contact with the substances and stimuli that keep us healthy? Through modern life, we are torn out of our functional circle with nature. It seems evident that separation from nature plays a central role in the development of these diseases.

Think about the effect of tree terpenes in the forest on our immune system. I proposed a change of perspective earlier in this book: In the woods, we don't have *more* natural killer cells and anticancer proteins in the blood than we "normally" do. Rather, we have *less* of these healthy substances in our blood, living in the modern world, than we would under natural conditions. We should consider the same change in perspective with DHEA. City life and modern living lead to *less* production of heart-protecting substances and a more rapid decline over the course of life than what would correspond to *Homo sapiens* in its natural habitat. This seems like a logical conclusion. Unfortunately, it cannot be found in medical books yet. There is no mention of the effects on the health of modern humans when we separate ourselves from nature. Everyone is well aware that we are exposed to environmental toxins, but the question of what we *lack* from not being in nature seems to interest only a few scientists in medical research.

In addition to effects already mentioned, DHEA is linked to a number of other health benefits. This steroid hormone supports the treatment of some mental disorders, including schizophrenia, a disorder in which perception and coherent thinking are disturbed.[4] Some schizophrenia patients hear voices or see things that are not there.

They often have a greatly reduced motivational drive. Schizophrenia has nothing to do with the multiple personality disorder in which the ego is divided into several personalities. The fact that many people today believe that schizophrenia patients are "several persons in one" is a blatant error caused by the misrepresentation of schizophrenia in movies. DHEA has also proved to be therapeutically effective against the severe form of depression called major depressive disorder.

In laboratory tests, DHEA showed favorable effects on cell cultures representing Alzheimer's disease. The hormone prevented the breakdown of fat molecules in brain cells—a process that leads to the progression of Alzheimer's disease—in petri dishes.[5] Doctors are now following this clue. There are also studies that clearly demonstrate the positive effects of DHEA on human muscle, bone, and sex functions.[6] The antiaging effect is still being debated among scientists.

When I was combing through studies on the health-promoting effects of DHEA, I noticed that most authors see the potential of the versatile adrenal gland hormone less in the administration of medical preparations than in a "hormonal optimization." In other words, DHEA is unlikely to have a future as a pharmaceutical pill against the mentioned symptoms, but scientists are looking for ways to stimulate the body to increase production of the hormone on its own. This is exactly what time spent in forests and other ecosystems achieves in a completely natural way free of side effects. Once again, this underlines the principles of nature-human medicine. It's a matter of strengthening the body's natural protecting and healing mechanisms—that is, a rebalancing of the human body and its functions that have been negatively affected by harmful environmental influences and the lack of contact with nature.

As we have just seen, a single hormone such as DHEA affects a variety of organs and functions of the body—and even the psyche. Here, the complexity of the human organism becomes clear. If you turn a single screw, the whole system changes. Adding an ingredient can have a chain reaction of effects. The opposite applies as well; removing an influence can result in positive or negative effects. Since the different subsystems of our body are interwoven, researchers—we can say with certainty—are still at the beginning of a great journey to decipher the complexity of human beings.

The fact that we still know very little about the effects and inter-actions of certain substances in our body is the reason many medical experts and biologists warn against, for example, the use of pesticides in the agricultural industry. You might have already heard of the "cocktail effect." Even if we assume that the thresholds, which are often arbitrarily and partly determined in the interests of industry, are appropriate for certain agrochemicals and that the individual substances do not pose a risk to human health as long as their concentration in foods does not exceed those thresholds, a *combination* of different chemicals can none-theless pose a considerable risk. We know almost nothing about how pesticides lumped together affect us and what potentially harmful inter-actions they cause in our organs. This cocktail effect can also occur with other environmental toxins that we do not absorb through our food. In this modern age, we could easily modify Paracelsus's well-known saying from "The dose makes the poison" to "The sum makes the poison."

Forests also produce a cocktail effect, but that one is good for us. The healthy effects of nature may be due to the sum of the many largely unexplored natural substances of which we have recently deciphered only a few. Therefore, isolated tree terpenes in laboratory experiments may cause positive changes in the immune system, but every study demonstrates a stronger effect in the woods. Forest air is a cocktail of bioactive substances.

The functional circle of humans and nature is highly complex, so much so that we will never decipher all its connections. We can par-tially attribute the fact that the adrenal cortex produces more DHEA in the forest to the tree terpenes that boost our immune system at the same time. Forests and natural landscapes also affect our psyche and our nervous system through stimuli and sensations, and not just through a biochemical exchange. The increased release of heart-protecting substances might also be related to psychological and neurological mechanisms that are activated in nature.

We have known for a long time that humans are a psychosomatic entity. The term "psychosomatic" is often misunderstood in common usage to indicate a psychosomatic illness is something imagined and doesn't "really" exist. But this concept is completely wrong. Psychosomatic symptoms are always "real" symptoms that are actually present and that we physically perceive. The only difference is that they can't be blamed entirely on physical causes. For example, mental and chronic stress can lead to inflammatory processes, gastrointestinal discomfort, or immune deficiency because psychological processes manifest themselves physically through the nervous and hormonal systems. And vice versa, physical processes influence our mental condition.

Psychosomatics is the medical science of the unity between psyche and soma, which means "mind and body" in Greek. Our mental experience cannot be separated from our physical processes. They depend on each other and are linked together. The often-used metaphor of "two sides of the same coin" is another misunderstanding. Mind and body are not two sides. The two sides of a coin are separated from one another along the middle axis of the coin. They face two different directions, and we can only ever look at one side at a time. But psyche and body are *one*. Such a clear distinction, such as between the two sides of a coin, would be impossible between physical and mental processes.

That being said, psychosomatics is a separate medical and biological science in which humans are referred to as "psychosomatic entities." I propose we expand this concept by one word: a person is an "*eco-*psychosomatic entity." The concept of eco-psychosomatic medicine has not been given due attention in medical science. I hope I can begin to change that with this book.

As demonstrated, our biological functions constantly transcend the boundaries of our body. Through the functional circle with nature, the psyche and body can no longer be separated from our environment, as they cannot be separated from each other. We even have organic "antennae" in us. Through this pipeline, the healing code of nature deeply affects our organs and cells, as we shall see shortly. Eco-psychosomatics shows the close connection between mind, body, and nature. So far, I have used the term "nature-human medicine." From this chapter on, I will refer to it as "eco-psychosomatics" instead.

## Organic Antennae and Their Network in the Body

A German journalist asked me once where the parasympathetic nervous system (PSNS) is actually located in the human body. I had previously told him that it was responsible for stimulating rest activities and played an important role in deciphering the healing code of nature. This sounds like a single nerve, which is found in a certain place in the body, but the PSNS is actually a complex neural network that is distributed throughout our entire body. These nerves of rest emerge from the brain stem as well as several parts of the spinal cord into the rest of the body. They belong to the autonomic nervous system, which controls many body functions. The exit points of the PSNS from the spinal cord are mainly found in the lowest segments, which are called "sacral segments," where the sacrum is located.

The network of nerves from the parasympathetic nervous system is connected to the eyes and ears via neural pathways. It also maintains connections to the lungs, bronchi, heart, stomach, and intestines, as well as to cells of the lacrimal glands, salivary glands, and the pancreas. Furthermore, the PSNS is connected to the kidneys, bladder, ureter, and urethra—to the entire urinary system. All of these connections run through nerve tracts, which enter the organs and penetrate the tissue as tiny branches in order to reach the cells. The network of the parasympathetic nervous system makes the connection of different organ systems in widely distant areas of the body possible. Thanks to the PSNS, our organs can react to what the eyes see and the ears hear. It is a gateway for perceptions from our environment, specifically from nature. It is an "organic antenna" that distributes its network of information into our innards. I chose the word "antenna" because there is a connection to the outside that enables a reaction to environmental influences. And I used the word "organic" because the parasympathetic nervous system is a biological system, but above all, because it spreads its network to the cells of our organs.

Why do some scientists call it the "rest and digest" system? This is because it is responsible for relaxation. It is active when we are not in danger, don't feel stress, and don't need to concentrate. In short, more electrical impulses pass through our network of the rest and digest nervous system when we feel safe and relaxed.

Let's recall that the nervous system of rest is part of our autonomic nervous system that regulates the body's unconscious actions. It is not alone in doing this. Like the parasympathetic network, the sympathetic nervous system is also part of this autonomic nervous system. The two are counterparts, so let's call the sympathetic one the nervous system of "excitement." It is also connected with our organs. Although the nervous system of rest and the nervous system of excitement work as antagonists, they do not do so in the way that one tries to dominate the other. Similar to the opposing Chinese symbols yin and yang, the two nervous system networks make one common, functional unity. They are both involved in the third part of the autonomic nervous system, namely, the enteric nervous system, also called the "abdominal brain." A Wikipedia article on yin and yang says, "Yin and Yang rise and fall alternately. After a high phase of the yang, a decrease of yang and an increase of yin follows."[7] The opposing forces of the system of rest and the system of excitement can be described similarly. Ideally, the two are balanced; however, this ideal is not always achieved. The sympathetic nervous system (the system of excitement) is active when we are attentive, tense, or in fight-or-flight mode. In stressful situations, its neurons begin to fire more. When we no longer have to pay close attention or a dangerous situation has passed, the activity of the system of excitement subsides again, and the opposing system, the parasympathetic nervous system, boots up again.

For all living beings, it was necessary in the course of evolution to develop a regulating system that determines whether we can relax or not. Both the nervous system of rest and the nervous system of excitement are connected with our sensory organs. They are, as I've said, our antennae to the outside. But they are also linked to the brain via the central nervous system. Deep in our skull lies an evolutionary part of our brain that biologists call the "reptilian brain." As the name suggests, we are not the only ones carrying it around. Many other animals have it, even reptiles, to which we are only distantly related.

In the reptilian brain—its evolutionary age is 500 million years—is the brain stem. Right above this, the limbic system sits pretty much in the center of our brain. The limbic system is also referred to as the "old mammalian brain" and is about 250 million years old. And on top of the reptilian brain and the limbic system is the cerebrum, also known as the "new mammalian" brain.

The reptilian brain, with the support of the limbic system, is responsible for controlling unconscious body functions. For this reason, it is closely linked to the nervous system of rest. Most of the nerves from the system of rest originate directly in the reptilian brain. It also has access to the nervous system of excitement via the spinal cord. Depending on the sensory impressions from outside, the reptilian brain and the limbic system head for one of two opposites—either yin, the system of rest, or yang, the system of excitement. The reptilian brain and the limbic system are the entities that decode outside influences—that is, the codes from the environment. Then they quickly decide whether we are in a situation in which we have to be highly attentive and maybe even run away, or whether we can relax.

You can imagine the types of stimuli that trigger these ancient brain systems to activate the nervous system of excitement. For Stone Age people, it was the sight of a mammoth or a predatory animal. It was also the necessity of balancing on a rock near an abyss during the hunt. Hunger and thirst were also dangerous situations in which arousal and active food search were necessary. Relaxation is placed on the back burner in these situations. Barren landscapes, such as deserts, that promise no nourishment or water are nature impressions that activate the nervous system of excitement rather than the system of rest. The same applies to dark or hidden areas in the landscape. Behind dense undergrowth or an obstacle could lurk danger. "A case for the nervous system of excitement," thinks the reptilian brain and limbic system, and they still function this way today.

The nervous system of rest, on the other hand, puts the organism in relaxation mode and makes it possible to focus on things and activities that do not serve basic survival. This includes creative work, too. The system of rest initiates modes of regeneration and growth. Then the focus is on healing, cell renewal, recovery, and enjoyment.

The following table shows the effects of the parasympathetic and sympathetic nervous systems on our organs and organ systems. Take the time to briefly review the comparisons and let them sink in before you continue reading.

## EFFECTS OF THE PARASYMPATHETIC AND SYMPATHETIC NERVOUS SYSTEMS ON OUR ORGANS[8]

|  | Sympathetic Nervous System (Excitement) | Parasympathetic Nervous System (Rest) |
| --- | --- | --- |
| Overall body | Puts the body in alert mode and prepares for flight, fight, or intensive food search; increases attention | Puts the body in relaxation, healing, and regeneration mode and enables activities that are not only for basic survival, such as creative work |
| Heart | Speeds up heart rate | Slows down heart rate |
| Blood vessels | Constricts blood vessels | Expands blood vessels |
| Blood pressure | Raises blood pressure (because of the constricted blood vessels) | Lowers blood pressure (because of the expanded blood vessels) |
| Lungs | Dilates bronchi in the respiratory tract | Constricts bronchi in the respiratory tract |
| Stomach | Inhibits digestion | Stimulates digestion |
| Intestines | Constricts blood vessels in the intestines | Expands blood vessels in the intestines |

| | Sympathetic Nervous System (Excitement) | Parasympathetic Nervous System (Rest) |
|---|---|---|
| Kidneys | Stimulates the release of stress hormones adrenaline, norepinephrine, and cortisol in the adrenal glands and reduces urination | Reduces the release of stress hormones adrenaline, norepinephrine, and cortisol and increases urination |
| Eyes | Dilates pupils | Constricts pupils |
| Salivary and lachrimal glands | Inhibits salivation and secretion of tears | Stimulates salivation and secretion of tears |
| Liver | Boosts glucose production | Normalizes glucose production |
| Blood | Raises blood sugar levels (because of increased glucose production in the liver) | Lowers blood sugar levels |
| Pancreas | Reduces production of insulin and digestive enzymes | Increases production of insulin and digestive enzymes |
| Sex organs | Stimulates orgasm when sexually aroused | Builds up sexual arousal |

The effects of the system of rest and of the system of excitement are in and of themselves regarded as neutral. It would be wrong to say that the activity of one is "healthy" and that of the other "unhealthy." Depending on the situation, human beings and other mammals need both nervous system networks to adjust the body to specific environments or situations. The interplay and equilibrium between parasympathetic rest and sympathetic excitement enable a balanced functioning of our body whose organ systems must be regulated to align with environmental influences.

Neither the neuronal fire in the system of excitement nor the impulses in the system of rest are better or worse than the other; we need both. However, here comes the big "but."

## Reptilian Brains in the Modern Age

If the nervous system of rest and the nervous system of excitement—as with yin and yang—are in a natural balance, neither of them makes us sick. It is evident from the previous table, however, that there are clear parallels between the activity of the sympathetic nervous system (SNS) and modern lifestyle diseases. Constricted blood vessels, increased heart rate, and elevated blood pressure occur when the system of excitement is active—the same phenomena that lead to cardiovascular disease, coronary heart disease, and heart attacks. The system of excitement inhibits digestion, which is very helpful in a dangerous situation when we need all of our energy for our extremities in order to escape or defend ourselves. That is when digestion must wait and not consume any of our body's energy. But there is another parallel to the health status of modern *Homo sapiens*: many of us suffer from chronic digestive weaknesses and problems in the gastrointestinal tract. Frequently, no organic cause can be found, so medical experts speak of "irritable bowel syndrome" or "nervous stomach." For those affected, these diagnoses are not particularly helpful since, of course, there is no conclusive treatment for these diseases that have causes that are not yet fully understood.

The activity of the SNS causes the liver to increase the production of glucose, raising the blood glucose to levels needed in stressful situations. At the same time, the pancreas reduces its insulin production.

Insulin is a hormone that lowers the blood glucose levels. Many people's blood glucose levels today are too high. Diabetes is a widespread lifestyle disease. This shows another parallel between the system of excitement and common lifestyle diseases.

The problem with the nervous system of excitement is that it is not gauged for the environment we live in today. It comes from a completely different, archaic world. Our reptilian brain can work wonderfully with the codes of nature. In modern life, however, it is flooded by stimuli that do *not* come from nature. Neurobiological studies have shown that the nervous system of excitement is overworked and switched on too frequently. This leads to its dominance over the nervous system of rest, and the harmony of yin and yang is disturbed.

It is not just environmental stimuli that lead us to a long-lasting state of stress. Although concrete wastelands, pollutants, traffic, noise, light pollution, and other potential dangers unknown to our reptilian brain play a role in everyday life, we should also take a look at our social structures when searching for causes of stress. The human mind and body are not suited for many of our professions, and social stress in the workplace can also trigger the stress reactions of our nervous system. This is because our SNS responds to *any* danger that threatens our existence. Today, our abilities as horticulturists and farmers are far from essential for our survival. Ironically enough, the amount of money in our paychecks is what matters now. Many people—I dare say, without a scientific reference for once—would change or even leave their workplace if they and their family did not depend on the income from their job. Pressure at the job, onerous working conditions, and anxieties regarding our material existence can lead to an activation of the SNS just as easily as predators can because we could also be risking "life and limb" within the framework that prevails in today's society.

Most of us know that unpleasant feeling of opening the mail when we are expecting a large bill. The minute we become aware of a menacingly high sum, a physically noticeable activation of the SNS and its networks kicks in, with all the mentioned symptoms. Many people begin to sweat. Perspiration is a typical accompanying phenomenon when the neurons fire in the sympathetic nervous system. Evolutionary biologists interpret sweaty palms as a preparation for flight because our ancestors could then hold on tighter to tree branches. Unfortunately, this well-meant effort on behalf of our sympathetic nervous system (to help us escape from danger) is no longer useful today.

Even though we know that outstanding bills are just a matter of numbers on pieces of paper, it does not help us any, since these numbers actually equate to an existential threat in the world in which we live. When it comes to unpaid bills, companies are not afraid to shut off the electricity, cut off the heat, or initiate other measures that can place our physical or mental health at risk. From this point of view, our reptilian brain is by no means wrong when it puts us into alarm mode via our SNS. It's just that the triggered changes in our body do not help solve these problems. But the neurons don't know it—they act according to their old wiring. Pressure to perform or bullying at school or work can also trigger those reactions.

In stressful situations, the adrenal glands emit more stress hormones such as adrenaline, norepinephrine, and cortisol, which inform the rest of the body that it is in an alarming situation. This phenomena is also due to the activity of the sympathetic nervous system. Stress reactions of short duration are harmless. It is only harmful when the stress persists for a long time—that is, if it becomes chronic. The attack of a lion comes to an end. If the attacked person survives, the stress reaction will also fade away. The same applies after paying bills. It may be "painful," but afterward the matter is taken care of. However, if a problem takes on such dimensions that there is no end in sight, the stress response never goes away, or it is continuously reactivated.

Imagine that a collection agency is putting you in a very uncomfortable situation. Yes, some agencies literally "persecute" people. Debt collection agencies do not care about your nervous system of excitement or about your health. They are not even interested in the origination of the bills they are supposed to collect. Where did the invoice come from? Did it really come about in a fair way? Does the collecting of the bill possibly destroy someone's financial existence while the invoicing party is a millionaire? In general, these kinds of questions are not asked by the collector. Imagine that you can find no way out of the agency's persecution. With each letter from the collection agency, you would be reminded of the threat, and each time, your reptilian brain would put you in alarm mode because your existence is—absurdly enough—*actually* in danger. The chronic stress reaction is in full swing.

People in these situations often suffer from constant perspiration and nervousness, sleep disorders, cardiovascular problems, digestive disorders, depression, difficulty concentrating, and so on. Often their mental health and ability to function at work decreases, which can worsen the situation, leaving them caught in a vicious circle. All of these are symptoms that are due to the activity of the sympathetic nervous system. This biological system is simply not made for the dangers of the twenty-first century!

We know from numerous medical studies that this kind of chronic stress makes us sick. Furthermore, the stress hormone cortisol has a direct, negative effect on our organs. It shortens our life expectancy, destroys brain cells, and can cause diabetes.[9] The overstimulation of the SNS because of psychosocial and physical influences in the industrial world is one of the reasons for an increase in lifestyle disorders, which includes mental disorders. Psycho-oncology is the science of the human psyche associated with cancer. Psycho-oncologists agree that chronic stress can be one of the causes of cancer development because it directly affects and weakens cells of the immune system.

For the health of modern humans, it is therefore beneficial to find ways to reduce the activity of the nervous system of excitement. This is done by strengthening its natural counterpart, the nervous system of rest. I wrote about the parasympathetic nervous system (PSNS) at the beginning of these observations. After a short detour into the problems of modern society, I now return to this system. Remember that the parasympathetic nervous system is directly connected with our sensory organs. Its nerves emerge from the reptilian brain and spinal cord, and it extends its network of rest deeply into our organs. That's why I called it an "organic antenna." The PSNS is able to "see" plants and animals as we move through nature. It is what translates the healing code of nature into the language of our organs.

## How Nature Makes Us Healthy Using Our Organic Antennae

The parasympathetic nervous system of rest is able to counteract the hyperactivity of its antagonist, the nervous system of excitement. In field studies, scientists at Chiba University in Japan and the Chungnam National University in South Korea were able to clearly measure how spending time in nature activates the PSNS. The strongest effect was seen in areas with a sparse number of trees, similar to the savanna, followed by forested areas where parasympathetic activation was also evident. The investigations were carried out in thirty-five regions throughout Japan. Spending time in the city had no comparable effect.[10]

The scientists were very careful to ensure that the activity of all test subjects was the same. During the multiday study, the participants slept in comparable hotels and received the same food. That way, researchers were able to rule out that the effects were due to the type of food the subjects consumed during the study.

The healing code of nature significantly activates the network of the PSNS when we see trees and other views of nature, hear birds or the rippling of a creek, spend time among trees, and even look at pictures of nature or listen to recordings. But what does it mean when the nervous system of rest is activated?

In the table on pages 82–83, we compared the effects of the two opposing parts of our nervous system. Roughly speaking, we can see that the system of rest normalizes all organ functions that the system of excitement placed into a state of emergency. The PSNS slows the heartbeat back down, dilates the blood vessels, and lowers the blood pressure. A research team at Nippon Medical School in Tokyo found that these positive effects can be seen after walks in nature, but not after walks in the city.[11] Spending time in green spaces, therefore, has the potential to counteract cardiovascular disease and lowers the blood pressure in hypertension patients. The nervous system of rest offers the necessary portals of entry, an organic antenna, through which the healing code of nature can affect our organs and cells.

The researchers at Nippon Medical School documented that spending time in a forest lowered the blood glucose level among people with diabetes. This effect was due to the forest and not due to medication. This does not mean that this disease can be cured by eco-psychosomatics, but it does mean that nature can make a contribution to the treatment of diabetes and protect us preventively against it. This is not surprising, and it is easy to explain. In the woods, the system of rest regulates the production of glucose in the liver and at the same time stimulates the pancreas to deliver more insulin into the blood. This hormone, which lowers the blood glucose level, is also administered as a drug for diabetes patients.

As already mentioned, forest medicine field studies have shown that the amounts of the stress hormones adrenaline, norepinephrine, and cortisol in the blood decrease significantly when people spend time in a forest or areas sparsely populated with trees. Here, too, it is easy to see how this is related to the nervous system of rest. It reduces the release of these stress hormones in the adrenal glands. The fact that contact with nature can help prevent digestion problems is also due to the parasympathetic nervous system. In particular, gastric and intestinal troubles without explicable causes, such as irritable bowel syndrome, can be alleviated by experiencing nature. The explanation is that the PSNS activates digestion, provides better circulation in the intestines, and stimulates the production of digestive enzymes in the pancreas. These are substances that help to break down our food into smaller particles so that it can be absorbed by the body.

I mentioned earlier that time spent in nature also sparks our creativity. As an example, I wrote about Michael Jackson's "giving tree," which has indirectly influenced music culture. It is easy to comprehend how stimuli from nature activate the parasympathetic nervous system, which stimulates our imagination. The system of rest signals to our entire body that there is no danger at present and that we do not have to be prepared for fight or flight. This opens the door to daydreaming and creativity, for these are manifestations of life that are not essential for basic survival. These activities become harder when we are in the alarm mode and easier when the parasympathetic resting network is active.

Neurobiological investigations have shown that a mere image or sound recording of trees, forests, and birds set the nervous system of rest in motion. We can also bring the healing code of nature into our home or to our sickbed. In many clinics, patients are already offered relaxation therapies accompanied by birdsong, forest and water sounds, or images of nature. Rachel Kaplan, the professor of environmental psychology at the University of Michigan I quoted earlier, was able to show that even brief glances out the window can reduce stress in the workplace and increase our ability to concentrate, as long as the view out the window is of green spaces or trees.[12] These effects are due largely to the activation of the nervous system of rest. It doesn't work if we can only see other buildings through the window.

In a study conducted by the University of Michigan with more than one hundred participating schools, those schools that had windows with a view of nature had better results on standardized psychological performance tests regarding learning and concentration and, in particular, had a lower rate of school dropout. In addition, the schools with green surroundings produced more future college graduates than other schools.[13] Apparently, a view of nature makes learning and attending school more enjoyable. A view of green space incidentally encourages even the problem-solving ability of pupils and social cohesion among classmates.[14]

At the beginning of this book, I reported on a clinical study published by the health researcher Roger Ulrich in *Science* in 1984. Recall that during several years of research, Ulrich demonstrated that patients recover faster and have less pain after surgery when their view from a hospital window is of a tree and not of a brick wall. But how is it possible that the mere sight of a tree is measurably affecting the self-healing powers of the human body and leads to positive effects? It couldn't be caused by the terpenes that the tree emits because they can't penetrate the window to enter a hospital room. Ulrich also excluded other influences in his carefully designed study. The determinant was the mere view of a tree. Based on our observations so far, we can now put the pieces of the puzzle together as follows: Complex interactions occur between our nervous system and our immune system. Our mind, nervous system, and immune system are inextricably interwoven. This is another piece of evidence that our psyche and body are not two sides of a coin but rather form a unity. Any change in a part of this system will result in changes in the others.

While stress activates the nervous system of excitement, which then temporarily restrains the immune system to save energy, the nervous system of rest sets the immune system back into full swing. Having surgery and lying in a hospital are stress factors. The nervous system of excitement fires away. The view of the side of a building does not activate the system of rest, as we know from the experimental comparisons between urban and forest walks. Views of nature, on the other hand, can. The PSNS was more active in "tree group" patients than in "brick wall group" patients. Its positive influence on the immune system accelerates healing and reduces the need for medication. The immune system fights off infections at the surgical site and promotes repair of the body tissue. Patients can return home sooner and have less pain.

After careful consideration, the strong effect of the tree view is not surprising. We've simply become somewhat desensitized to the vast connections between humans and nature. This is why Ulrich's study results seem amazing at first glance. They really are not anything out of the ordinary. We are natural beings, and our species has been in a functional circle with nature from the beginning. Over eons, the *Homo sapiens* nervous system has processed countless perceptions of nature as inner images. This is how visual codes emerged, which our body understands. The healthy effect of spending time in a forest comes, therefore, not only from chemical substances like terpenes but also from visual impressions in the woods.

Jakob Johann von Uexküll, the man who coined the term "functional circle," said in the nineteenth century that impressions from the outside are processed as inner signals and symbols. Sounds, pictures, and other stimuli from nature have become important to the human organism only gradually over the course of evolution. It is not solely due to random copying errors in our genome that we have learned to decipher the code of nature; it is also due to the experiences of our ancestors and their active interactions with nature. We assess certain stimuli from the world of plants and animals as soothing and positive, not dangerous, because they were important for the survival of our ancestors. The singing of birds in the treetops is not interpreted as a danger; instead, it creates a soothing ambiance. The same applies to the rippling of a small creek that trickles down a grassy slope. We do not judge these phenomena of nature as threatening, but rather we associate them with security and the availability of vital resources. We have internalized this over eons.

I already mentioned that studies have shown how we respond to certain tree species particularly well. We intuitively consider those trees attractive that offered our ancestors protection from predators or intense sunlight in the savanna or those that supplied them with food. This effect continues today with modern people. Flowers often produce edible fruits. Stately mushrooms, edible berries, the black elderberry fruits, and the bright rowanberries in autumn captivate the attention of nature-loving people. To our eyes, they stand out from the overall vegetation cover, and we feel attracted to them. Everything edible and beneficial in nature is very well known to our biophilia. Our biophilia is calibrated to it.

Field studies have shown that people in all climate zones have a strong preference for calm and sparkling water surfaces. This is also a legacy of our ancestors. Once it was important for us to recognize drinking water in open landscapes even at a great distance. The light show on the water surface signaled the availability of vital resources from afar. Without our millions of years of history, it would not be possible for us to classify different impressions of nature intuitively and effortlessly into "dangerous" and "reassuring."

The following landscape elements are especially effective at activating the nervous system of rest:

- Standing and sparkling water surfaces such as ponds, lakes, and lagoons

- Calm flowing waters

- Seas and oceans

- Flowers, blooming trees

- Gardens with fruit trees and vegetables

- Berry bushes

- Calm spots in the woods where we can see or smell mushrooms

- Birds and birdsong

- Trees, especially those with rough bark, a knotted trunk, or an expansive canopy that provides ample shade

- Clearings or meadows with scattered trees, bushes, or groups of trees (savanna-like landscapes) [15]

I would like to leave no doubt that there are also sensations from nature that can trigger stress reactions and have a negative impact on our health. However, this does not in any way reduce the value of the positive nature stimuli for our overall health.

## How Economic Demands Are Destroying the Organs of *Homo Sapiens*

Before I change the subject, I'd like to continue a little with the topic of our nervous system. We have dealt extensively with the two great opposing forces of our nerves. In the end, we focused on the parasympathetic nervous system because it is relevant to the healing effects of nature. It has also become clear that our modern lifestyle leads to an overactive sympathetic nervous system, and the balance between both systems is disturbed. We have seen how this has facilitated the development of lifestyle diseases, and I will shed more light on one aspect of this. I would now like to present additional evidence that our organs may be permanently damaged if we live under the requirements of our economic system—and that the healing code of nature can distance us from the damage caused by the economy's demands.

Let's recall the example of the collection agency. I have made it clear how the ruthlessness of the system in which many people must live, or rather "function," can result in permanent activation of our body's stress system because financial problems, unpaid bills, or difficulties at work are perceived as existential threats. Under the conditions that prevail in our society, such problems are *indeed* life-threatening. The modern requirements of our economy do not take human health into account. Workers are pressured into increasingly longer weekly work hours. Pressure to perform is increasing, and solidarity among workers is becoming more and more difficult since many people are afraid of standing up against injustices in the workplace. They are at risk of losing their income, and this alone—within the established rules of the game—decides whether a person loses their "existence" or not. The conditions at many work-places and training programs are sufficient for an overactivation and, especially, for a continual reactivation of the system of excitement.

This stress reaction causes the adrenal glands to release more of the stress hormone cortisol into the blood. It then reaches the brain through the bloodstream, where it binds to the neurons and triggers overactivity via special receptors. In the cell nucleus of the neurons, certain gene sequences are activated and read, which contain the blue-print for proteins that we need in greater amounts when under stress. The overactive neurons allow more calcium to penetrate through their cell membranes. Neurobiologists are not yet sure precisely how this mechanism works. However, the increased activity of the brain cells is probably due to the fact that our body assumes that we can think our way out of the stressful situation.

This is clever, but there is something that our circuits do not know. It's not so easy to throw a wrench in the works of our eco-nomic and social systems—and thinking about it doesn't help. Many people's hands are tied because they cannot easily change jobs and are dependent on their employer. Therefore, they might have to continue putting up with a tyrant or rising pressure to perform. Others cannot break free from the claws of bankruptcy or foreclosure and must help-lessly watch as everything is taken away. And because these situations don't change, the nervous system of excitement is constantly cranked up and triggers the same processes in the brain again and again. The brain cells continuously absorb more calcium than in a normal state.

When neurons are overloaded with calcium, they simply die. They cannot withstand the recurring overstimulation. Based on the latest findings, neurobiologists believe that persistent stress at work or in strained social situations can lead to permanent damage to the human brain. This could explain the increase in mental illness in modern societies.

Observations made by behavioral biologists with baboons support the hypothesis that chronic stress reactions can damage the brain and even lead to physical illness and death. Mental stresses are translated into physical manifestations through circuits in the body and present themselves as organic diseases. The baboons that the behavioral researchers examined had been kept in cages. The scientists discovered that this stressful situation led to severe consequences for these primates: they developed stomach ulcers, inflammatory bowels, and damaged kidneys; their neurons died in large quantities and left behind organic brain damage; and many of the baboons died.

Relevant to this sad event is that these primates did not get sick because of the cruel and objectionable conditions of the cages or the bad diet, but because of the social stress caused by their situation. Due to the limited space, the baboons could not live out their intrinsic behavior and develop a natural social structure. In the wild, the weaker male baboons avoid the dominant ones. This was not possible in the cages. The ones lower in the hierarchy were always at the mercy of the higher ones, and it was mainly the lower ranks that became sick and died. Biologists performed autopsies on the primates and were able to show by means of the physical findings that *social* stress was the actual cause of this misery.[16]

We humans, like the baboons, are primates. That is why these observations are important for us. When I read the results of this study, I inevitably had to think about the situation of many people in the modern working world. Even if the pressures and problems at the workplace become intolerable, it is often very difficult to escape the hierarchical structures and social stress. The correlation between this predicament and the increase of physical as well as psychological lifestyle diseases is obvious, and science is gathering evidence for it. But will something change in our society? Will the working world ever be "species appropriate"—that is, fair—for *Homo sapiens*?

At the moment, it's not looking good because people seem to be considered more as a "resource" than as human beings. The demands of cost-effectiveness and profit dehumanize the world we live in more and more, no matter how often our pitiful nervous system of excitement fires away. This type of health hazard transcends all forms of danger against which our body has developed strategies over the course of our evolution. But at this point, nature comes into play again, as we turn to another healing service from the ecosystem in which we live.

## Being Away: "I'm Off for a Bit, Then"

Environmental psychologists call another pleasant effect of nature, which every human being has probably already experienced, "being away." This term was coined by psychologist couple Rachel Kaplan and Steven Kaplan, both professors at the University of Michigan. Their work is internationally known among environmental psychologists and doctors trained in environmental medicine.

Henry David Thoreau, an American author and philosopher who lived from 1817 to 1862 and was one of the founders of American Romanticism, wrote, "When I would recreate myself, I seek the darkest woods, the thickest and most interminable and, to the citizen, most dismal, swamp. I enter a swamp as a sacred place,—a *sanctum sanctorum*. There is the strength, the marrow, of Nature."[17] Thoreau is regarded as a pioneer of modern ecological science through his precise descriptions of the ecological balance of lakes. He gained his insights by intense observations of the sixty-five-acre Walden Pond in Massachusetts, as well as other natural lakes. He described in detail the movement of shorelines and groundwater levels, the vegetation in and around the lakes, and the changes of water currents throughout the year. With his observations, Thoreau built the foundation of our understanding of natural ecological cycles of freshwater lakes.

When we are in the wilderness or in natural, unspoiled landscapes, we are able to gain distance from societal problems. This is what environmental psychologists mean by the term "being away." Trees, rivers, and birds do not send us bills. They don't unleash a collection agency on us. In the wilderness, we are one living being among many. Animals and plants do not condemn us for our appearance or attitude. It is still possible to experience wilderness even in our industrially dominated modern age. There, where the land cannot be used for commercial purposes, where no factories can be built and, above all, no endless agricultural deserts can be created, we still have some nature reserves, which are perhaps no longer virgin but are still basically a wilderness. Think of mountain ranges, marshland, rocky landscapes, and so on.

Sitting by a high mountain lake and watching the water dance while an eagle circles high in the sky gives us noticeable distance from our society, which means distance from social or professional problems as well. We know, for example, that bills are on the table at home, but up in the mountains, we experience another reality in which these bills have no power over us. They are nothing more than printed pieces of paper. We know that once we return to civilization, we will be confronted with the old problems, but during our time in the wilderness, we succeed in gaining a distance that is healing. Numerous studies have shown that we are not simply repressing our problems. Because nature inspires us for new approaches, new ways of thinking, and looking at problems, it helps us find new ways of problem-solving. I write about this in detail in my book *The Biophilia Effect*. In order to provide a basis for our further discussion here, I will briefly summarize the most important examples of these studies.

The environmental sociologist Angela Meyer at the University of Montana carried out extensive studies in the Rocky Mountains in order to show that while people are staying in the wilderness, they actually gain distance from social burdens and worries and profit from this for a long time.[18] She chose women to participate who, because of their sexual orientation, experienced bullying at the workplace and in social environments. All participating women were lesbians, bisexual, or transsexual. They slept in a research station in the Rocky

Mountain wilderness for several days where scientists study forest ecology as well as nature-human relationships. The women went on hikes, bathed in natural waters, and even enjoyed the experience of showering under a waterfall.

After the women left the wilderness, Angela Meyer discovered, through science-based interviews and questionnaires, that the nature experience was of great significance for the women. They all reported that they felt free from other people's judgments. Even though they knew that the prejudices in our society continued, they were able to dissociate themselves from it—they were "away." In doing so, they experienced a world in which their sexual orientation didn't matter. The women reported feeling accepted and said that the natural environment had also helped them to feel less inhibited in social exchanges with one another as well as to be supportive of one another. Of course, the women eventually returned to their old surroundings, where nothing had changed. What had changed, however, were their own attitudes.

Angela Meyer's investigations were social science experiments. All the studies I have described so far were based on measurable values and statistics. Scientists call this "quantitative" research because solid figures are what count. Angela Meyer, on the other hand, chose a "qualitative" approach for her wilderness studies. Qualitative research is used more frequently among sociologists than among other scientists and physicians. It is not a matter of investigating, for example, the effectiveness of tree terpenes against cancer by comparing large groups. It is about the state of an individual. How does the healing code of nature affect this woman or man? What does she or he experience? What changed in his life or in her health after the wilderness experience? In other words, it is about the *quality* of the experience. In this research approach, scientists always remain close to the participants and gather the results by means of personal interviews and written questionnaires. In addition to laboratory methods and clinical studies, the healing code of nature can be explored in this way also.

Angela Meyer's investigations showed that after spending time in the wilderness, women reported increased self-confidence and greater freedom from the opinions of other people. They traced these positive changes back to their experiences in nature.

In all likelihood, it would be necessary to take regular outings in nature in order for these effects to become permanent. The popularity of a pilgrimage—for example, along the Camino de Santiago in Spain, which leads through remote natural landscapes—might be related to the state of "being away," being away from social, professional, and existential problems and being away from a condemning society in which we must function like a cog in a wheel.

The US Department of Agriculture released a comprehensive scientific publication with the title *Wilderness Visitor Experiences: Progress in Research and Management.*[19] It includes a large number of studies that researchers from various universities conducted in the Rocky Mountains. The results are unmistakable: wilderness experience is an aid in dealing with social crises and helps with existential needs. The nervous system of rest kicks into action and prevents damage to our health. This does not solve our problems, but it does protect us from getting sick because of them. In addition, there is an effect that scientists call the "immediate conscious experience" (ICE) in the wilderness. ICE is all the conscious content that arises in a human being while spending time in nature. Wilderness studies show that nature inspires us to see and think in new ways about social and professional crises. Wilderness experiences often lead to new problem-solving strategies. This was demonstrated in several of the studies published by the US Department of Agriculture.

As I said, animals, plants, boulders, and rivers that we encounter in nature won't change our society or the way people treat each other there, but they can change *us*. In the wilderness, we experience that there is a reality on this planet where economic requirements, professional performance, or bills from collectors are just as unimportant as our appearance, physique, or sexual orientation. And this reality is greater than humankind's social systems. The world of plants, animals, fungi, stones, rivers, and algae will survive eons longer than all the factories, accounting departments, and economic systems in the world. What reality could provide us with more distance from it all than nature?

In addition, symbolism in nature is important to the immediate conscious experience. Picture, for example, boulders gradually being conquered by plants until soil forms somewhere on the formerly naked surface; trees might even appear and, in an ideal situation, a forest might arise. Or imagine a tree that has fallen after a heavy storm and is now sprouting new roots because its vital forces do not give up. Nature is full of symbols for growth and new beginnings, for the permanence of life. It is precisely in situations of crisis that these symbols, and the conscious experience they trigger, give us strength. Let us not forget that we, too, are beings of nature, and these vital forces are equally effective in us.

Psychologists refer to "coping" when they talk about psychological and social management strategies. There is no doubt that "being away" in nature is a very effective coping strategy for many people.

Because of its diverse, positive effects on the human mind and body, wilderness experience is already being used in therapeutic interventions—for example, in the treatment of alcohol and drug addiction, social phobias, or anxiety and panic disorders.[20] This is eco-psychosomatic therapy.

## Forest or Savanna?

Different natural landscapes also contain different healing codes; spending time in a forest is different than in a savanna. I use the term "savanna" in a somewhat broad sense, namely, for a type of landscape that we find almost everywhere on earth. Savanna-like landscapes are green spaces with sparsely scattered trees and bushes or with small clumps of trees. Our developed landscapes often take this form. Fruit orchards or pastures scattered with groves also fall under the savanna-like landscape definition. Many urban green spaces, such as Central Park in New York and Hyde Park in London, are designed as a savanna. This is due to the fact that landscape architects, who also study aesthetics and environmental psychology during their training, have known for decades that sparse woods are particularly soothing to us and promote mental regeneration.

Evolutionary biologists explain this "savanna effect" with our tribal history. I have already mentioned that the transition from *Homo erectus* to *Homo sapiens* took place in the East African savanna. Our intuitive preference for the savanna landscape and the types of trees that grow there shows that we are still shaped by evolution and carry inside us a collective memory of our common past in this landscape. But there is also a second explanation for the relaxing effect of savanna-like landscapes. We can easily survey our surroundings if they're sparsely populated with trees. Because we can see through the trees and bushes, our archaic warning systems don't think there are hidden dangers or attackers.

We intuitively look to forest glades or meadows with trees for places to relax. This does not mean, however, that the positive effects of being in a forest pale in comparison to the benefits of the savanna. Because of its healthy air, saturated with tree terpenes, the forest is a place with the biochemical healing code of nature. The activation of the PSNS can also be observed in the forest, and the stress hormones disappear. The heart-protecting substance DHEA becomes more active, and our immune system is stimulated to form more defenses. Nevertheless, the anti-stress effect is even higher in areas with sparsely populated trees than in dense forests.

When jogging in the forest, I sometimes notice that I feel different in dark areas than in a light-flooded part of the woods. As I move through the thicket, where I cannot see far and there is little light, my running speed always increases. This happens automatically, and when I leave these areas, I feel relieved. That is when I realize that my reptilian brain has led me through a supposed danger and gone into alarm mode.

Sometimes I jog through tree plantations, where spruce trees are growing close together with almost no sunlight between them. These are artificial forests and would never grow this way naturally. They feel monotonous and lifeless. Because no light penetrates to the ground, almost no herbs or grasses grow there. The earth is covered by a uniform layer of brown needles. Such gloomy tree plantations, which are not natural ecosystems, seem to especially disturb my reptilian brain. When I run through this area, I reach my peak running speed because my energy is redirected to my legs to help me get away quickly. Clearly, my nervous system of excitement is most active there, and the system of rest remains dormant.

However, most of the forests I exercise in have a very calming effect on me. They are usually mixed forests with a multilayered tree canopy. It is true that these kinds of forests are not as open as savannas, but sufficient light shines through. Most forests in the lower elevations are like this. To me, they are soothing and inspiring.

Forests also remind me of my childhood. My grandfather was a forester and a botanist and regularly took me on walks through the woods, where he often told me stories and fairy tales. Together, we imagined how fairies or other mythical beings were waiting for us behind tree stumps or in magical glades.

The forest has many effects on us. When I was reading Jean-Jacques Rousseau's work *The Confessions* in the summer of 2013, I always sat under the same beech tree in the woods. I could not have imagined a better place to read this book. Rousseau's autobiography is full of captured nature moods from the eighteenth century. The forest atmosphere around me underlined the words of the great Swiss-French poet and philosopher. I felt completely safe and relaxed at this place in the forest. Whenever I hear the name Rousseau, I inevitably remember "my" beech tree.

I am convinced that the atmosphere of the forest has always attracted people and has inspired stories, fairy tales, poems, and songs.

Let me continue with Rousseau for a moment. In his autobiography, the philosopher reported how nature inspired him to paint during his walks through the forest. "I dispose of Nature in its entirety as its lord and master; my heart, roaming from object to object, mingles and identifies itself with those which soothe it, wraps itself up in charming fancies, and is intoxicated with delicious sensations. If, in order to render them permanent, I amuse myself by describing them by myself, what vigorous outlines, what fresh coloring, what power of expression I give them!"[21]

**THIS CHAPTER IN A NUTSHELL**

Our search for explanations about the healing effects of nature has led us to a hormone from the adrenal cortex called dehydroepiandrosterone (DHEA). DHEA is an endogenous heart-protecting substance. The level of this steroid hormone in human blood decreases over the course of life, which is one of the numerous biological reasons for the aging process. Medical studies have shown that spending time in nature, but not time spent in the city, leads to a significant increase in DHEA production.

Clinical studies have shown that DHEA counteracts numerous cardiovascular diseases. It protects us, among other things, from dangerous coronary heart disease in which the heart muscle is not sufficiently supplied with blood because of loss of elasticity in the blood vessels. Doctors have additionally discovered the therapeutic efficacy of DHEA in the treatment of mental illness and Alzheimer's disease.

Our nervous system of "rest and digest" acts as a kind of organic antenna and entryway for the healing code of nature, which can reach and affect our organ cells this way. The nervous system of rest, the parasympathetic nervous system (PSNS), is a large network of conductors and neurons that run from our cranium and spinal cord into almost all parts of our body and is connected with our organs and cells. The counterpart of the nervous system of rest is the nervous system of excitement, or the sympathetic nervous system (SNS). The two together are like yin and yang in forming the overall nervous system and are supposed to remain balanced. In stressful situations, the system of excitement is activated. Bodily functions such as digestion, immune defense, and so on slow down. Our heart rate, blood pressure, and blood glucose levels rise. Social problems, existential angst, and pressure at work, as well as the absence of sensory stimuli from nature, can lead to the dominance of the SNS (excitement) over the PSNS (rest). Through complicated regulatory circuits, this can lead to common lifestyle diseases, including cancer. Recent studies have shown that overactivation of the nervous system of excitement in humans and other primates can cause serious organic damage, brain damage, and death.

Due to evolution, nature is full of healing codes that activate the counterpart of the nervous system of excitement, the nervous system of rest. This counteracts physical and mental lifestyle diseases and activates the mode of cellular healing.

Through nature and wilderness experiences, we also gain distance from societal, professional, and financial problems and can regenerate during this time away.

# 5

# THE HEALING BOND BETWEEN HUMANS AND ANIMALS

## The Time I Saved a Forest Animal

On a rainy day in August 2015, I was driving on a highway in a region in Austria known as the Waldviertel (Forest Quarter). As the name suggests, this area is heavily wooded. On a highway bridge, I saw on the left shoulder of the road, directly in front of the guardrail, a 15-inch-tall figure. It was crouched dangerously close to the fast lane of the curve. I looked over while driving by and only had a second to see what it was. Just a moment later I was already past it.

"Was that an owl?" I asked myself. My immediate impression was that it was a stately, owl-like bird and that it had stuck its head back into its feathers. That was a bad sign. It must have been hit by a car. I wanted to check on this bird, but the next exit was still three or four miles away. I took that exit, looped back, and drove past the bird again. It was still there and had not moved from the spot. "Damn, I can't stop anywhere here." I drove past the creature one more time, and this time I knew it was not going to fly away because it seemed to be stunned.

I took the exit again and stopped at a gas station. There I rushed into the shop and asked for an old cardboard box that was big enough for the endangered bird. I took a look in my trunk. My work gloves were there, so I had everything I needed to pick up the bird.

I drove back toward the highway bridge, but this time from a side road. I hoped to get to the median strip next to the bridge. From there, over a hill, there was a way to reach the place where the bird crouched. But it was hopeless. There was heavy traffic, and any effort to cross the road to get to the median was life-threatening. Fog was limiting everyone's vision, and the danger of not being seen by a driver was too great. But I did not want to abandon the animal, so I contacted the police and asked for an escort.

I am sure that not every police officer would have been as helpful and flexible as the two I ran into. We had arranged to meet at a parking lot near the bridge. When the patrol car arrived, one of the officers got out slowly and moved toward me with deliberate steps.

"Why do you think the bird needs help?" He asked in an unfriendly tone.

I explained to him that the creature had been sitting on the edge of a highway bridge for at least forty minutes, with its head pressed down into its feathers, and hadn't moved.

The officer said he and his colleague were willing to escort me to the bridge, where I could stop and cross the road. "We will cover your back, but we are explicitly not liable. If you do this, it's entirely at your own risk," the officer explained sternly.

I agreed, and we drove in a convoy to the highway.

As we had discussed, I stopped at the place where the bird was sitting. The police car stopped behind me. On the officer's signal, I crossed the highway and walked about sixty-five feet along the bridge to the place where the bird was. The fog had turned to rain, and I arrived there dripping wet, but it was an exhilarating feeling to be near the bird at last. It sat with its back to me, so I could reach through the guardrail unnoticed. I positioned my hands left and right over its shoulders and grabbed it with lightning speed. The feeble bird of prey spread its wings that spanned about three feet. I could feel its wet, heavy plumage under my thick gloves and saw that there was a considerable amount of blood coming out of its neck and eyes.

It was a tawny owl, which is a medium-sized bird. I closed the box and ran back to the car. One of the officers approached the car door. "Can I see it?" he asked. The sharp tone from before had completely disappeared. I opened the box. The stately tawny owl was beautiful despite its injuries. Satisfied with what he saw, the official nodded to me and said, "I hope it'll make it."

Half an hour later, I was sitting in the waiting room of a veterinarian who is well versed in wild animals and treats them free of charge. The doctor and I already knew each other, and we had something in common. We were both friends of the forest and wild animals. When the door to the treatment room opened and the veterinarian came out, I was prepared for the worst, but the news wasn't so bad, considering the circumstances. The doctor had seen the owl immediately, and it was sleeping now, I was informed. I was led into a room that was used as an intensive care unit for animals.

The tawny owl was lying in a glass oxygen chamber and was attached to tubes. It rested on its right side, and its plumage rose and sank with each difficult, yet powerful breath. An X-ray image hung next to the glass chamber. The veterinarian switched on the light, and the whole skeleton of the bird was visible in the picture.

"By the way, it's a male owl. The good news is that he has no broken bones," the doctor said, "which is remarkable because he must have been hit head-on by a car, if not a truck."

I breathed a sigh of relief. "And what is the bad news?" I asked.

"The owl has pretty much every internal injury it could have. There is a risk that it will bleed to death internally. Blood is even coming out of its eyes. I gave him a coagulant that stops internal bleeding. We can't do much more at the moment, and surgery is not possible with so many internal injuries."

I stayed with the bird for a while and felt glad that the police and I had succeeded in rescuing it.

While this being lay in front of me in the glass oxygen cage, I felt tremendous respect for it. Tawny owls are mysterious animals. Even as a child, I was thrilled by them. Through the window of my childhood bedroom in the greenbelt around my hometown of Graz, Austria, I could look directly into a forest. In the summer, at night, I heard the calls of the tawny owls. The sounds reached my ears as if from another world. I could tell them apart and heard how they communicated with each other. The imposing calls of the tawny owls sounded mysterious. They echoed through the dark forest for a long time and captured my imagination at night. I loved this sound very much as a child.

The next day the veterinarian called me, "He woke up!" We were both happy. "He is standing again, if only on one leg. He can keep his balance when nudged, but he keeps his eyes closed all the time."

This bird was clearly tough. It had survived a serious accident on the highway, and then persevered there in the rain. At the veterinary hospital, it survived anesthesia during X-rays. And now it was apparently over the worst of it. However, its recovery process was still far from complete.

We took him to the bird sanctuary in Haringsee near Vienna, where they put him in a large aviary with a group of the same species. His condition became more and more stable.

"We may be able to release him back into the wild," the vet at the bird sanctuary said. "But if he doesn't start to fly again, or if it turns out that he is blind in one eye because of the accident—something we can't rule out yet—he will remain here in this group." Only one thing mattered to me then: He was alive!

The behavior of the police officers was exemplary. What fascinated me most was the enthusiasm and dedication of the two, despite their initial skepticism. I also found the efforts of the veterinarian commendable. He fought for the life of the bird, sparing no effort or cost. Solidarity between humans and animals does exist; however, our relationship to animals is divided. We lock up animals in "animal factories" to fatten them up in unnatural and inhumane ways. In slaughterhouses, they are massacred by contract workers. I have seen all of this myself during years of research for my earlier books. Even the organic sector is responsible for this inhumane treatment of agricultural animals. The meat is neatly packaged in the refrigerated display cases of the supermarkets—the cheaper, the better. Most consumers do not make a connection to the animal's past when they see its body parts prepared for the kitchen. However, when people have personal encounters with animals, most are fascinated. The two police officers were enthusiastic about the wild bird once they were involved with it. They even inquired about its condition and were delighted to hear that it had survived.

In animals we meet a "you"—there is no doubt about that. Phenomena such as factory farming or the breeding of foxes and minks in tiny cages for the "production" of their fur for coats are only possible because these creatures get lost in an anonymous mass of what we call "farm animals." In a face-to-face encounter between humans and animals, however, it is clear that animals are our relatives. Our modern societal relationship with them is very problematic.

When I visited "my" tawny owl at the bird sanctuary, I passed an aviary where a group of Ural owls was kept. These large birds used to call the vast forests located in Germany, Austria, and the Czech Republic home. Now the Ural owls I saw were being prepared for life in the wilderness so that they could be resettled in those forests.

I approached carefully. The heavyset birds' size is impressive; they stand between twenty-three and twenty-seven inches tall. As with all owls, they have wide shoulders and a flat face with a downward-pointing beak. One of the birds sat regally on a branch where I could see its profile. When it noticed me, it turned its head slowly and looked into my eyes. We looked at each other for a long while. My breath slowed. I had never been so close to such a rare bird before. My biologist's heart beat faster. Above all, I was overwhelmed by the expression it seemed to have on its face and the profound black eyes with which the bird looked at me. This being is a sentient life-form, *just like me*, a "person."

After a while, it let out a screech, turned around, and spread its wings. Slowly and elegantly, it glided into the rear area of the large outdoor enclosure, its wingspan almost five feet, and alighted in front of a wooden hut. The owl turned back toward me and hit its upper and lower beak together two or three times, creating a strange clattering sound as if someone were striking two wooden blocks together. The sound echoed over the landscape. I immediately realized why mysterious owls from the forest are in so many fairy tales and fables. These fascinating birds are indeed fabulous.

I wish I had had a neurobiologist with me that day. I would have taken a blood sample and measured my nervous functions during my encounter with the Ural owl. I bet there would have been physical differences before and after. And I would not be surprised if the changes had a positive health potential. I could deliver the proof of this in a later book! Fortunately, however, other scientists have already carried out such measurements. It turns out that encounters with animals have the same healing potential for humans as encounters with plants do.

## About Humans and Other Animals

At the beginning of my biology studies, I was tortured by a dissecting course. Every student received various dead animals throughout the semester that had to be cut apart properly. The professor demonstrated the procedure, using a video camera to project his "craftsmanship" on a monitor so that we could all see close-ups of what he was doing. Dissecting large exotic cockroaches was relatively easy for all of us to handle, even though I knew that insects are highly developed animals. When we got to work on a fish, it was our first vertebrate under the knife. I remember how it was to work on the brain of the fish. I felt as if I were entering a holy sanctuary of this dead being. After all, its brain had not only controlled important body functions, but had also been involved in maintaining the fish's awareness when it was alive.

One day, when representatives of the species *Rattus norvegicus* lay on the dissection table, it was hard for some of us to participate. Rats are, just as we humans are, mammals. Their internal organs look like ours, only smaller. Their fur is biologically identical to human hair and body hair. Rats have hair! "Our" rats had white hair.

The professor pointed out to us that dissecting mammals can be emotionally taxing because they are our relatives. You could feel the inhibitions in the air when we had to cut open their bellies to penetrate the internal organs. The heart looked like a miniature human heart. When we removed the hearts, some students dealt with their mixed feelings through humor. When one of the rat hearts flew out of its connective tissue in a high arc and fell to the ground because of a student's carelessness, some in the classroom laughed from an uncomfortable embarrassment. During the break, almost everyone agreed that dissecting rats was emotionally difficult. In the afternoon, joking and laughing continued—clearly a stress management strategy. When we finally opened the skull and exposed the brain, everyone in the room went silent. Nobody laughed anymore out of embarrassment. It was quiet.

Rats have a mammalian brain built like ours, with a brain stem that fulfills the same functions. Remember, this is the 500-million-year-old "reptilian brain," which we share with reptiles and amphibians. In rats as well as humans, nerves of the nervous system of rest come from the brain stem and spread into the organs as a network. *Rattus norvegicus'* limbic system is located above the brain stem, as in the case of *Homo sapiens*. This is the "old mammalian brain," which is involved in the creation of emotions in all mammals, including humans.

When the rats were dissected, we saw the cerebrum of these animals, which is connected to the brain stem and limbic system. The cerebral cortex is called the "new mammalian brain," exactly the same as with humans. Although our cerebrum is naturally much larger than a rats', we can ascribe intelligence and consciousness to rats. I mention this even though I assume that no one doubts it these days. Rats are able to learn, as numerous scientific behavioral observations have clearly demonstrated. Recent research shows that they are even more intelligent than biologists once thought. Anne Churchland, a professor of neurobiology at the University of California, San Francisco, has shown in her studies that rats are able to make decisions similar to humans by weighing different possibilities. The rats Anne Churchland observed were not only able to understand light signals as guides, their brain is also capable of more complex decisions, and experiments have shown that when they are presented with tasks, these animals are as likely to make the best possible decisions as are humans.[1]

At the University of Georgia in Athens, on the Oconee River, researchers found that rats are capable of self-reflection, just as primates are. They can gauge their own abilities and include this information in their decision-making. They know what they can do, and how they are overstrained, for example, when trying to overcome obstacles in order to get food.[2] Rats also support each other to solve problems through a combined effort.

Anne Churchland summarizes the topic of her research at the University of California in the following way: "Animals are faced with many decisions. They must integrate information from a variety of sources—sensory inputs like smell and sound as well as memories and inner impulses—to arrive at a single behavioral output."[3]

In other mammals, such as sheep and goats, not to mention humanlike primates, the "new mammalian brain" (the cerebral cortex) is a much larger part of the brain than it is in rats. The brains of such mammals look a lot like our own "gray matter."

When the dissection class began to dissect larger animals, I no longer participated. After the rat, I dropped the course. The rats were laboratory animals from some experiments that I didn't know about. Even back then I doubted the necessity and meaningfulness of many of those experiments. Therefore, I decided I did not want to dissect any more killed creatures. I didn't see the benefit of this experience when I had on my bookshelves excellent textbooks that depict every detail of anatomy faithfully and clearly. In addition, I was able to look at the department's three-dimensional anatomical models of animals and organs. Continuing to dissect animals would have given me no extra knowledge. I found a way to graduate without the course.

Around the time of these events, I ran into an old friend while riding the bus. Just before the beginning of my biology studies, she had started studying law. Our two majors did not have much in common. When I told her about the dissection class, she asked me if we humans were "animals." I replied that from the point of view of biology, we are considered mammals in the systematics of animals. With the term "systematics," biologists mean a complicated system of classification of living creatures, which is based on scientific, and now mainly molecular biological findings, and is constantly updated. I have never heard from the mouth of a serious biologist the idea of singling out humans in the systematics of the animal kingdom.

Our closest relatives, the chimpanzees and the bonobos, have around 95 to 99 percent of the same genetic code as humans, depending on the measuring method. The biological differences between humans and chimpanzees depend less on differences in the genetic material than on differences in the way the genes are read. For example, the human organism converts more of its genome into proteins in the brain than the chimpanzee does. Biologists do not yet know why this happens. What is certain, however, is that we are closely related to chimpanzees and bonobos as well as to other apes.

There is no biological dividing line between us and other animals; this boundary was artificially created. As I have already explained, our brain stem goes all the way back to reptiles and amphibians. When we trace back the evolution of life on earth, we land a few billion years in the past at the primeval cell LUCA. As you already know, LUCA is the last universal common ancestor of all earthly creatures, including humans.

When I explained this to my friend on the bus, a woman sitting across from us began to shake her head wildly. She gave me a withering look. Then she turned away from us, crossed her arms in front of her body, snorted, and looked out the window. I noticed a similar scenario on German television. On a talk show called *Nachtcafé* ("Night Café") on July 3, 2015, some vegetarians and nonvegetarians discussed the dignity of animals and the question of whether, from an ethical point of view, we should eat them, or more specifically, make them into commercial products. Guests on the show included the German sports journalist and meat enthusiast Waldemar Hartmann, the German actress and vegetarian Marion Kracht, and the Austrian artist Chris Moser, who is an animal rights activist and hopes for an end to factory farming. He told the audience that his motto is that art must always have a message. His sculptures and other artwork would therefore always raise questions about human rights and the dignity of animals. The discussion led to an argument between Moser and Hartmann.

The presenter said to Chris Moser, "Many have grown up with the biblically anchored position of 'Man must subdue the earth.' I think this is exactly what you're telling us is no longer the case.

Mankind should not be allowed to subdue the animal. Is this a correct interpretation?"

Chris Moser replied, "Yes, without the biblical aspect, but basically, I see humans as one of many animal species, and no one species holds a higher position over others. This means, of course, that humans and animals have to work side by side." Just as humans should not oppress each other, they should not oppress animals, the artist continued.

While Moser was still replying, a decided "No way" was heard from Hartmann. Hartmann expressed his rejection of Moser's words with a dismissive hand gesture and said loudly and firmly, "That's when I stop listening! So if we're just one of many animal species—great, but . . ." He finished his rebuttal with a patronizing head shake.

Humans are sometimes offended when someone insinuates that our species is related to other animals. Charles Darwin, the founder of the theory of evolution, was confronted with this behavior in the nineteenth century. Snarky caricatures in newspapers at that time showed Darwin's head on a monkey's body. *He* might be a descendant of monkeys, but *we* were not—that was the message. Darwin's contemporaries were mistaken then about something that is still a common misconception today. The human being does not come from the apes, but we humans and today's apes have a close *common ancestor* that lived in trees.

The close relationship between humans and other primates is obvious if you think once again about our genetic match with chimpanzees of 95 to 99 percent. The social life of most apes is also very similar to a human social life, especially regarding their societal structure. Chimpanzees and gorillas use simple tools in a strikingly similar way to humans. While writing these lines, I put a nut between my teeth. The way I hold the nut, the movement of my hand, and my lips taking the nut—the chimpanzees, bonobos, gorillas, and orangutans would do this exactly the same way. Our bite and our digestive tract are very similar to those of the apes. These similarities are, of course, based on common ancestry and not by chance.

On the basis of genetic and molecular biological findings, *Homo sapiens* is categorized in the biological system within the primates under the "dry-nosed" primates (haplorhines). Nearly all biologists agree on this classification. Humans and animals are side by side in zoology textbooks, and most general zoology lectures at universities include the subject of the human organism. Zoologists are often the ones who research and teach the physiology of human beings at medical universities—that is, the biological functioning of our organs. One of the world's leading academic textbooks for human anatomy and physiology is *Anatomy & Physiology: The Unity of Form and Function*.[4] I have two editions of this commendable tome on my bookshelf. In the United States, this is the anatomy book most used for training physicians. The author is Kenneth Saladin—a zoologist!

Throughout my career as a biologist, I have never felt degraded or humiliated as a human being because of the facts I learned about my biological proximity to other species. It is not an insult to be a creature of nature and to come from the same network of life as the other animals. What is so bad about that? The vehement shaking of the head with which some people react to our position in nature can only be explained by the assumption that these people have a negative attitude toward animals. Anyone who sees a dignified counterpart in animals, a "you" that they can *meet*, cannot possibly feel bothered when scientific facts about the human-animal relationship are mentioned in their presence.

But if we recognize that all creatures are related and the human is positioned among the primates, does this mean we're not supposed to make distinctions between different species? This is what people like Chris Moser—people who consistently regard *Homo sapiens* as a species among many and not as the "crown of creation"—are frequently accused of. I don't know anyone—not a zoologist nor an animal protection advocate—who seriously doubts that there are *differences* between the species. It is not about being the same. Life has been developing in countless directions for billions of years. Some forms of life branched off from our lineage a long, long time ago, or we from theirs. The evolutionary family tree of all creatures is extremely branched and unimaginably complex. Some animals have conquered habitats like the deep sea or Arctic tundra and adapted to these environments. They obviously have very little in common with us because their functional circle with the nature in which they are embedded is quite different from our habitat. With primates, however, we have remarkable similarities, so much so that many zoologists and philosophers even demand that primates be given certain basic rights. This includes the right to freedom for our closest relatives and thus the end of captivity. This would also put an end to animal experiments on chimpanzees and others.

Based on the position of these scientists, another basic right that should be granted to primates is the right to be left alone by humankind. Wild primates should not be pursued, injured, or killed by us. That would mean the end to hunting primates. I expressly endorse these demands for concrete reasons. The degree of consciousness of chimpanzees, bonobos, gorillas, and orangutans, as well as other primates, is definitely close to that of humans. Primates are aware of their own existence. They have an ego-consciousness. They are able to recognize that other living beings also have an "I." Countless studies have demonstrated that apes can learn our language. They are able to understand simple sentences and answer questions using sign language. They can learn basic grammar. Primates are able to feel mental suffering. Locking up, torturing, and killing apes should be judged as a serious matter from a bioethical point of view.

The fact that humans are currently the species with the most complex culture is undisputed. This does not mean, however, that other animal species are not capable of consciousness and psychological suffering. Being self-aware has evolved continuously in the course of evolution. There is no place in the animal kingdom where we could draw a line to classify all beings above it as conscious and all others as unconscious.

Whoever uses the word "person" when referring to animals will be going out on a limb, just as the artist Chris Moser did on *Nachtcafé*. Only humans are able to be persons, people say again and again. This may be legally true, but I am a biologist and not a lawyer, so I am about to go out on a limb.

I considered the tawny owl as "someone" during the rescue operation on the highway. The Ural owl that looked into my eyes at the bird sanctuary also had a "personality," at least to me. *Someone* with *personality* is a person: this combination is completely allowed in logical terms of language, entirely independent of the law. Behavioral biologists have long recognized the existence of individual personalities in vertebrates and particularly in mammals. In biology, this is called "phenotypic behavioral expression," and this term is used for humans and animals alike. It means nothing more than "personality"—and this always results from a mixture of predisposition and social and environmental experiences.

In humans and other animals, behavioral researchers distinguish between two basic, very different types of personality. These are, if you will, the extreme poles between which personalities can move. *Proactive* people and animals face problems and challenges in life more quickly and more actively than the *reactive* ones. They are more dominant and quickly explore their surroundings and social environment, but they do it rather superficially. They quickly develop behavioral habits that they do not like to change. The *reactive* humans and animals are not dominant and respond less quickly to challenges. They are more reliant on others for problem-solving, so they behave more passively than the proactive ones. They need longer to explore their environment, but they do so in more depth.

There are five main characteristics that psychologists use as a guide to describe the personalities of human beings. These traits differ in degree. Psychologists call them the "Big Five." They can also be expressed and understood in the form of five questions:

1. How emotionally stable is the person?

2. How extroverted is the person?

3. How open is the person to the new?

4. How socially apt is the person?

5. How reliable is the person?

In psychology, these five fundamental questions help describe the human personality with meticulous mathematical precision, which might appear strange to a psychological layman like me, and perhaps also to you. A modern psychology major is almost a mathematics major. Behavioral biologists have found that the Big Five used by psychologists to describe human personalities are also suitable for describing the character of animals. For this reason, some researchers also use these cornerstones of personality psychology with animals now. Animals, like humans, can be more or less emotionally stable; they can be extroverted or introverted and more or less socially apt. Some of them are open to new experiences, while others prefer to always leave everything as it is. Animals can be reliable or less reliable in social coexistence. All of these possibilities that create personalities are not attributable only to *Homo sapiens*.

The boundary between human and animal is not a biological one, but rather has been artificially established by us. In nature, this distinction cannot be found. Still, I will continue to refer to "human" and "animal" because this has become firmly established in our linguistic usage.

## A Friendship with Animals

The previously mentioned evolutionary biologist Edward O. Wilson from Harvard University recognizes in humans a biologically oriented tendency to be drawn to living animals. Like plants, animals also express life and growth processes that our biophilia is looking for. We can only passively participate in communication with trees by absorbing their chemical communication molecules (terpenes) via our lungs, mucous membranes, and skin. With many animals, on the other hand, we can enter into an active communication process, which is even reciprocated. Anyone who has ever lived with a dog or a cat can confirm that a conscious communication with animals is possible.

Here is another biological commonality between humans and animals: all vertebrates exhibit similar functional areas in the brain, which behavioral biologists call a "social behavior network." This network is an important biological basis for social behavior. Anatomically, the social behavior network of vertebrates consists of six organic components, which are the same in all species. When I write about "vertebrates," I don't mean only humans and other mammals, but also birds and fish, as well as amphibians and reptiles, which are far removed from us from an evolutionary perspective. In all these vertebrates, including *Homo sapiens*, the same hormones are formed and the same mechanisms maintained in the social behavior network. Some biologists speak of an "evolutionary toolbox" of social behavior.

I do not want to say that our social coexistence is controlled by chemical processes in the brain and that we are something like "social machines." Our brain, however, provides the biological basis on which to build our personality and behavior. Personality development is, therefore, an active learning and shaping process for which we use the evolutionary tools that have been given to us. What is inside a toolbox and which tools are available to us is innate. How we use the tools and what we make with them is in our own hands.

Scientists established the hypothesis of the "social brain." The brain of humans and other animals is an organ that helps us to actively react to environmental conditions. On the other hand, it is also a social organ. Just as the liver is the organ of detoxification and the intestine is one of digestion, the brain is the organ of relationships to the environment and to other living things. The brain is an important biophilia organ.

In a textbook of behavioral biology that deals with the connection between humans and animals, I came across the following statement: "It is not surprising that within primates (including humans), cognitive skills are shared. Humans and chimpanzees split only four to six million years ago. However, birds and mammals split around 230 million years ago. Therefore, it is truly surprising that there is hardly any primate social phenomenon or cognitive skill that has not also been demonstrated in birds."[5]

I observed how familiar birds are with their caregivers at the bird sanctuary where I had taken the badly injured tawny owl. They recognized certain caregivers as they approached the enclosure and could distinguish them from each other. The birds maintained different relationships with different employees. The lead veterinarian told me that he had lived in biocenoses (communities) with different birds, including ravens, crows, jackdaws, and blackbirds. These birds lived a kind of "family life" with him, and he even let them inside his house. They flew completely freely but did not fly away from him. He told me about a crow that had lived with him. The crow even accompanied the veterinarian and his dog on bike rides and always flew beside him. If he stopped, it sat on his shoulder or on a branch and waited. The crow was jealous of the dog when the veterinarian would pet it. The bird would try to drive the dog away and then sit on its human friend's shoulder or head. The zoologist and behavior researcher Konrad Lorenz coined the term "imprinting" for when birds and other animals are socialized to live with humans. Lorenz gained his insights from living in close and intimate community with wild animals.

For a while, a blackbird lived with me and wandered freely around my house. She had fallen out of the nest as a young bird, and I raised her at home. When I sat and worked in front of my computer, she sat on my shoulder and looked at the screen. I have a photo of the two of us watching YouTube videos together.

This kind of friendship with a bird is not just trivial fun; it is a serious social connection. After this experience, I am sure that close friendships with birds can be a form of support in crisis situations. Cat and dog lovers often report that their animals help them in periods of sadness or even depression and that social interaction with them is a great enrichment in difficult times. Bird friends could have the same effect. The blackbird was seriously interested in me and clearly not just interested in being fed. When I set her free, she came to my front door every day for a long time and slept in a safe place in the garden shed. Later she met a male blackbird, and her visits became less frequent. The two of them settled in a nearby tree at first but later moved away.

Now I have a group of white-rumped munia, *Lonchura striata*, living in my home in a big aviary that I designed in a near-natural way. These birds, originally native to China, are the size of a tree sparrow and have beautiful brown-gray feathers with a bronze sheen. "My" white-rumped munia came from an animal shelter. These friendly birds are very social—at night, they always sleep together in a nest they built themselves, and they regularly maintain the nest collaboratively. As soon as a hole becomes visible, they fill it immediately with blades of grass. The entire construction has to be fixed again and again, requiring real teamwork.

The white-rumped munia males almost never quarrel, and I have never seen a territorial fight or a skirmish between males over a female. Once I had to take one of the two cocks from the aviary—this is what we call male birds in general, not only among chickens—and put it under quarantine because it was sick. I temporarily placed it in a cage in my music room. Contrary to my worries, it did not grow lonely but began to show immense interest in interacting with me. During this time, it became tame, which is very unusual for small songbirds. I realized that this bird loved music. Whenever I played an instrument in the room, it craned its neck in my direction and observed everything closely. It usually started singing along. White-rumped munia males are capable of a unique song, which is very complex and goes far beyond the warbling of canaries. This musical bird liked my instruments, the ukulele and piano, but most of all it liked my marimba: an African xylophone with rosewood bars and an enormously warm, full sound. Marimbas are also used in major orchestras. My simple marimba is a custom instrument, built entirely of wood, on which tones sound in a very natural way. For me, it is a real "biophilia instrument," precisely because such a basic instrument makes such beautiful music. I believe the little bird liked the instrument so much because of the primal sounds. In any case, there was a real connection between us, and the little guy even began to actively call for me.

Friendships between people and birds are not rare. "We observe here at the bird sanctuary that birds form close friendships with each other," the veterinarian told me. "These friendships also develop between different bird species." Birds are very social animals and able to establish relationships with other bird species and with humans.

For a long time, humans have maintained close relationships with other animal species, too, species that were not initially friendly. Today's pets and farm animals look back on a 10,000-year coevolution with us. Our ancestors influenced their development and in this way helped shape the changes in their genetic material and thus in their physique and behavior as well. By the same token, domestic animals have influenced our human evolution. The fact that dogs and humans get along so well is a result of this coevolution.

During my studies, I had a zoology professor who took us on numerous excursions to the mountains. There we observed wild animals and searched for their tracks in the landscape. The professor was always accompanied by his silver-gray husky. Except during the lectures, I never saw him without his companion. Sometimes the husky was even there during a lecture, and it definitely didn't miss a single excursion.

The professor and his dog often went off together when we were in the forest, while we were conducting outdoor zoological experiments that the professor had given us. I remember seeing the two of them on a precipice high up in the mountains. They were sitting on a boulder with their backs to me, looking out over the wide landscape beyond the peaks of other mountains. The wind blew through the gray, curly hair of the professor and through the silver-gray fur of the husky, which is biologically hair as well. The deep connection between man and dog was evident to me at this moment. They were real friends, companions.

Dogs are the oldest pets of humans. "The human-dog relationship may be viewed as being based on a historic association between two species with large, socially intelligent brains," according to Kurt Kotrschal, professor of behavioral biology at the University of Vienna, and his colleagues.[6] The two species that found each other are *Homo sapiens* and *Canis lupus*, the zoological name for the wolf. Earlier zoologists hypothesized that dogs formed by crossing coyotes and jackals with wolves, which our ancestors were supposed to have done over the course of thousands of years. Thanks to modern genetic research, however, we now know that dogs descended exclusively from wolves. Wolves are also the only wild animals with which dogs still voluntarily form communities today. The wolves take them into their packs and even reproduce with them. In regions where the wild wolf populations are heavily decimated due to human influences or excessive hunting, wolves will even partner with domestic dogs.

This "dogification" of wolves began about 30,000 to 40,000 years ago, long before humans began settling down as farmers. Thus, the dog is by far our oldest companion animal. The human-dog relationship originated in several regions of the world independently of each other and at different times. People of the Neolithic period began by capturing wild wolves and keeping them. Some evolutionary biologists believe that the wolves themselves hung out near humans to find food there and that some of them eventually lost their fear of humans.

Wolves that were later under the care of human communities were genetically isolated from their wild companions. As we know with the help of modern epigenetics, living beings can biologically pass on their experiences to their offspring relatively quickly. Tame "pet wolves" thus passed down their life experience with humans. The altered environmental conditions of the wolves were reflected in their genes since all organisms tend to adapt to their environment. From wolves gradually came dogs. That is why we can confidently speak of "dogification."

Through selection, and later by breeding, humans influenced the further genetic development of dogs. Our ancestors selected among dogs which would reproduce and which would not. This is basically a culturally shaped evolutionary process influenced by humans—the selection criteria were no longer dictated solely by nature. Dogs and all other pets have, therefore, a *biocultural human heritage*. They arose at the intersection between the history of nature and culture.

Of course, for a long time throughout human history, dogs were not only mankind's best friend, they were also used for herding sheep, for hunting, and as bodyguards. Skull findings show that 5,000 to 6,000 years ago, hunting dogs, which looked like mastiffs and greyhounds, lived in the ancient Orient.[7] The ancient Romans possessed a variety of hunting dogs, which were used in different forms of hunting. They harnessed dogs to pull their carriages, much like they did with horses, and they also used dogs to guard their houses, farms, and cattle. Today, there are hundreds of dog breeds. After thousands of years of mutual development, *Homo sapiens* and *Canis lupus familiaris* (the domestic dog) have become ideal companions. As the zoological name indicates, dogs are "family members" of humans.

*Felis catus*, another pet of humans, became a somewhat more independent companion than the dog. The domestic cat descended from the wild cat *Felis silvestris*, which was originally native to Eurasia and Africa. Cats, like dogs, have a strong social sense, and many of them are even capable of recognizing their human's feelings and reacting to them in a social manner. Making a distinction between animals as the "instinctive beings" and humans as the "social beings" has long since been dismissed in the study of behavioral biology. It is still a misconception in popular culture, though, that animals' empathetic behavior can always be explained by a person's observation error. In other words, animals only *seem* to be empathetic.

Imagine that the owner of a cat is depressed or sitting on the sofa crying. The cat comes and begins to rub up on its upset human's legs or "caress" her with its velvet paws. The skeptics claim that this cannot be a manifestation of empathy because animals are not in the position to know what is going on in another living being. Thus, the behavior of the cat could only be a sign of uncertainty, or a kind of "program" that runs to make sure the person will feed it when they are in a passive state and the cat fears that feeding time might be skipped. I have heard and read these interpretations so many times. From modern behavioral biology research, however, we know that social abilities, such as adopting someone else's perspective, have emerged over the course of evolution in many animals. Knowing the well-being of others is a survival skill of all social animals. Both in humans and in many other animal species, social skills have arisen from this necessity and have led to the formation of corresponding brain structures.

The now-famous mirror neurons are another reason for our innate capacity for empathy. We know that people with a disorder of this form of nerve cells have a diminished or lack of ability to relate to others, but this does not mean that empathy cannot be learned by these people.

If you believe that empathy is always associated with selfishness, because it is an evolutionary adaptation for survival in social structures, and that animals do not really act empathically when they turn to their caretakers sympathetically, but are actually acting for self-serving reasons, you will definitely find arguments for this position in evolutionary biology. But then to be consistent, you'd have to draw the same conclusion about human social abilities. Even in the case of humans, compassionate social behavior can be interpreted as the pursuit of a personal benefit, developed under the same evolutionary selection conditions as in animals. If you decide to subscribe to this theory, you have to apply the same standards to humans and animals. Then, compassionate behavior is not "real" compassion in us or in cats or dogs. In fact, this has been a heated debate among neuroscientists and biophilosophers for many years—a dispute about whether genuine compassion exists at all, even among humans.

Regardless of your position on this question, a distinction between "genuinely social" humans and "fake social" animals is not justifiable. In addition, cats, dogs, and many other mammals, just like we humans, are capable of learning new social behaviors. Ravens, crows, and about 120 other species from the crow family are capable of intelligent social contributions, such as those we otherwise only know of in primates. The same is also true of some species of parrots.[8] This has been demonstrated repeatedly in behavioral studies.

Empathy and even bonding are not only possible from person to person, cat to cat, or raven to raven, but beyond the species' boundaries. This too was often questioned not many years ago, but behavioral science has verified the healing potential that the connection between humans and animals has.

## Healing Encounters

In the first chapter, we began exploring the healing code of nature in the trees. The basis was scientific results that had demonstrated a clear link between the presence of trees and improved human health. The scientists were confronted with the question of how trees could have such an immensely positive effect on our body, and they found logical and comprehensible answers in the biology of humans and plants. Research about the healing bond between humans and animals began in a similar way. At the beginning, there was an astounding insight:

the presence of an animal, like the presence of a tree, has a positive effect on our immune system and our self-healing powers, and this can be scientifically measured.

In 2004, the international scientific journal *Psychological Reports* published the results of a research study conducted by health scientists on students. The test subjects were divided into three groups. The first group of participants petted a real dog, while the participants of the second group petted a toy dog. The students in the third group were asked to make themselves comfortable on a sofa and just relax. And, lo and behold, the measured immune parameters didn't change with the "sofa group" or the "toy-dog group." In the group, however, where participants had stroked a living dog, the researchers noted a significant increase of immunoglobulin A in the saliva. This is a form of antibody found mainly in body fluids such as saliva, mother's milk, and the fluids of the internal organs. There, they are among the most important of our antibodies. They provide protection against pathogens and prevent them from entering our body and cells through the skin and mucous membranes. They also form a barrier against intruders in the intestine.[9] The immunoglobulins play an important role in the promotion of healing the mucous membranes.

In numerous statistical studies conducted by scientists between 1980 and today, it was discovered that people living with a pet are on average healthier than other people of the same age with a similar lifestyle. The people with animals had lower blood pressure as well as fewer stress hormones in their blood and went to the doctor less often than people without pets.[10] Such studies were repeatedly conducted on large populations all over the world. An Australian study found that people with dogs and cats were prescribed fewer drugs for sleep disorders and fewer painkillers than people without animals. Social scientists Janelle Nimer and Brad Lundahl at the University of Utah in Salt Lake City demonstrated in a meta-analysis that medical and psychotherapeutic treatments are, on average, more effective when animals are involved.[11] Animal-assisted therapy has been implemented in many areas of health care.

Bruce Headey, a sociology professor and health scientist at the University of Melbourne, Australia, demonstrated in 1999 that people who live with dogs or cats are statistically less likely to suffer from cardiovascular disease.[12] Professor Headey's results were confirmed by other scientists. The biologist and medical scientist Erika Friedmann teaches and conducts research at the University of Maryland School of Nursing in Baltimore, where she is associate dean of research. Along with her colleagues, she clearly demonstrated in several scientific investigations that there is a statistical correlation between dogs and recovery from heart attacks in humans. If patients live with a dog after a heart attack, their chances of getting healthy again increase. Long-term survival rates are higher with a dog. Patients with heart disease are more likely to survive even after being disconnected from a heart-lung machine, and they will live longer thereafter if they have a dog at home. Some of Erika Friedmann's explosive study results were published by the world-renowned University of Cambridge.[13] Her studies also showed that social interaction with dogs, especially in older hypertension patients, reduces blood pressure.

Neurologist Andreas Zieger at the University of Oldenburg demonstrated in his clinical studies that social contact with rabbits, guinea pigs, and dogs leads to significantly better therapy results in severely brain-damaged patients and that even family members exhibit less stress in the therapy sessions with animals than in conventional therapy.[14] Zieger also successfully uses animals in treating persistent vegetative state (PVS) patients. His study results show that patient heart failure is significantly reduced by animal-assisted therapy. The physician observed a particularly marked decrease in anxiety symptoms. Levels of the stress hormone adrenaline can also be slightly reduced in PVS patients with animal contact. The health-promoting effects of rabbits, guinea pigs, and dogs are the strongest when they are allowed to be in bed with seriously ill patients in order to have body contact. It goes without saying that the highest hygienic standards are adhered to. Even just twelve minutes of daily contact with animals proved to be therapeutically effective in Zieger's studies.[15]

Similar results were obtained by neuropsychologist Stefanie Böttger at the Städtisches Klinikum München (Munich Municipal Hospital). She performed studies with hemispatial neglect patients, who no longer perceive half of their body and have very little control over it. Some hemispatial neglect patients even experience the affected side as not being part of their body, as if it were part of a stranger. In severe cases, patients lack any knowledge of having a disease, so they cannot be convinced that the body parts belong to them. Spatial perception is also limited. Those affected often do not perceive the left or the right field of vision. It is as if a part of them has "disappeared." In the conventional treatment of this neurological disorder, the patient is presented with numbers, symbols, letters, and pictures on the computer. They are encouraged to look for and recognize stimuli from the neglected side. At the same time, the computer program supports them, for example, with a grid in the background and flowing dots that move in the direction of the "disappeared" side to draw attention to it. The method is recognized as a standard therapy for hemispatial neglect.

Stefanie Böttger was able to show that animal-assisted therapy is better suited to treat this disorder. In her studies, she placed rabbits and other small animals in bed next to the patients to stimulate feeling, vision, and movement. After three weeks, she compared the results with conventional computer-assisted therapy. Both methods proved to be effective, although patients participated more actively in the animal-assisted therapy since animals are alive and provide an element of surprise. With the computer, on the other hand, patients seemed apathetic because of the monotonous nature of the stimulus. They themselves stated upon questioning that the animal-assisted therapy felt more effective and far less difficult than computer simulation and explained that it was because of the aliveness of the animals.[16]

All of these correlations between contact with animals and physical health are further evidence that the forces of biophilia are at work in humans—we notice this because people reach out to other life-forms—and that living out biophilia at all levels leads to measurable effects on our health. Animals should play an important role in eco-psychosomatics.

The positive effects of the human-animal encounter on mental health are just as well documented as the effects on physical health. Animal-assisted intervention is also used in psychiatric hospitals.

In a clinical study, physicians' treatment of a portion of their patients suffering from schizophrenia included animals for one year. These patients received animal-assisted psychotherapy four hours per week. A comparison group received just as much therapy but without animal contact. Tests showed that patients from the "animal group" had achieved a higher adaptive functioning level (the ability to take responsibility for oneself and to interact with others socially) at the end of the year than patients who were treated purely conventionally. These abilities can be severely limited in people with schizophrenia and make their lives considerably more difficult. Animal-assisted therapy helps with this psychiatric disorder better than conventional therapies without animals.[17] In another study, researchers tracked indicators such as the frequency of smiles and social contact with other people as well as visible signs of joy in schizophrenic patients. It, too, showed how the inclusion of animals in therapy resulted in patients smiling and interacting with other patients more frequently and expressing happiness more often and more intensely than patients from conventional therapy groups without animals.[18]

Contact with farm animals also triggers clearly measurable psychotherapeutic effects in people suffering from various mental illnesses.[19] Goats, sheep, horses, cattle, pigs, and so on are social animals with distinct personalities and are useful for therapeutic interaction with people. If patients in psychological distress are able to regularly be involved with and nurture farm animals, their health improves more quickly than if they were not given this opportunity. Depressive symptoms decrease, social inhibitions diminish, and new strategies are acquired for coping with problems.

Exposure to animals also helps in therapy for people with anxiety and panic disorders.[20] Scientists have known for a long time that self-efficacy (individuals' belief that they can execute behaviors necessary to achieve their goals) improves when animals are part of their therapy. Clinical studies investigating the psychotherapeutic benefits of farm animals have shown that regular contact is crucial for the success of the intervention. Long-term and repeated social situations with animals also lead to long-term improvements in mental health.

In our society, we consider these animals "livestock": animals that serve the sole purpose of producing meat and other products. The role farm animals play in the health of people in mental distress should change our view about these companions. The biological and social proximity of *Homo sapiens* to other species should not trigger head-shaking disapproval but rather provide a convincing opportunity to rethink our approach to other species. The intensive raising of livestock in factory farms runs its unfortunate course, even though the social abilities of these animals could help to heal and make people healthier.

Animal-assisted medicine and therapy are used in almost all areas of health care. The healing bond between humans and animals can be used therapeutically for all age groups. There are numerous studies that demonstrate how animals help in the treatment of children with autism spectrum disorder.[21] And instead of treating ADHD patients with Ritalin, some physicians and psychotherapists are successfully pursuing the path of animal-assisted therapy.[22] Researchers have demonstrated that measurable physical signs of stress in children during a dental appointment are significantly reduced when a dog accompanies them.[23] Therapy dogs are already living in some nursing homes and, with the staff, regularly visit residents in need of care. This form of therapy is gaining popularity, and there is a lot of evidence that animals have a measurable positive effect on the health and mental condition of seniors.

Some skeptics have argued that the contact with the therapy animal is not the healing agent, but rather the benefit comes from contact with the persons who accompany the animal or from other patients when they all play with the animal together. Researchers have shown that this is not the case. The positive effect of animal contact occurs when the patients are alone with the animal as well. In group situations, the therapeutic effect is actually slightly weaker than it is in individual situations. A study also shows that elderly people who care for a songbird or an aquarium are less depressed and have fewer symptoms of dementia than when they have no contact with animals.[24] Of course, family or nursing staff should help in caring for these animals so that the elderly are not overwhelmed with responsibilities.

Children who grow up with animals can interpret the nonverbal expression of emotions in other people better and more accurately than children who grow up without animals. Animal contact during childhood promotes social skills—which also benefit the adult later in life.[25]

Science has also provided clear evidence that an encounter with wild animals, provided they are not dangerous, is beneficial. Similar to the sight of trees, observing deer, squirrels, or wild birds lowers stress hormone levels in the blood, and the parasympathetic nervous system—the nervous system of rest—is activated.

That reminds me of the story of the musician Andreas Danzer, whom I interviewed for my book *The Biophilia Effect*. He is the son of Georg Danzer, a well-known "Austro-Rocker" in my home country. That is what we call musicians in Austria who bring a mixture of rock music and dialect singing to the stage. Andreas Danzer experienced firsthand the effects of the healing bond between humans and animals. When he was hospitalized for pulmonary tuberculosis for six months in 2011, he was not allowed to leave his room at first. He couldn't have left anyway because he was physically too weak. But once he received medical permission, he began to go to a nearby forest every day. He always sat on the same old tree stump at the edge of the forest. "There was always this family of deer," he said. "At first they kept a safe distance, but after one or two weeks, they accepted my presence and came closer. I sat right in the middle of them and felt like Dian Fossey in *Gorillas in the Mist*."

Andreas noticed that his depression due to illness decreased with every visit to the deer family in the forest. "I dared to hope again, and my strength to defeat the illness steadily grew. My fascination with the animals and the woods distracted me from my physical symptoms. The fresh air was good for my lungs, and moving helped build up my muscles again after spending so much time in a hospital bed. When I walked up the mountain to my spot, I sweated out the toxins from the medications, and the side effects decreased. I built up my physical and mental strength while a relationship between me and the deer family emerged."

He told me the animals motivated him to go on that daily hike, which was strenuous at first, because he did not want to miss a single evening when the graceful deer gathered in the clearing.

Andreas Danzer said he perceived himself as part of nature and part of the great cycle of life. He is certain that "everyone feels the need deep inside to be close to nature. We have roots, and they definitely did not grow in cement."[26] This is also an apt description of the healing code of nature.

Contact with animals leads to healthy changes in the concentration of neurotransmitters such as the "happy hormone" dopamine or the natural painkillers, endorphins. At Children's Hospital in San Diego, animals are used in the treatment of sick children. Researchers there conducted clinical trials and were able to demonstrate that the patients' pain at the pediatric hospital can be significantly reduced by touching a therapy dog, and that's after a mere twenty minutes.[27] Several clinical studies are currently underway to determine whether contact with animals can also relieve the pain in cancer patients. So far, the signs look good that this hope will be confirmed, and animals can be integrated into an eco-psychosomatic cancer treatment in the future.

## The Healthy Hormone Oxytocin

We have already talked about the social toolbox of the brain, which emerged over the course of evolution and which we can now access in order to actively make social connections with others. Friendships and relationships can also be formed between different species. This is possible because the social toolbox in the brain of almost all vertebrates contains astonishingly similar tools. I'd like to point out one of them because many of the health-promoting effects of human-animal contact are based on it: oxytocin.

Oxytocin is our bonding hormone and plays an important role in shaping social interactions with other people. For example, it is released in mothers and infants at birth and during breastfeeding. It is involved when we fall in love and supports the connection between two loving people on a hormonal basis. Today we know that oxytocin also strengthens the bond between siblings and friends.

What most people don't know is that the attachment theory, which psychologists have developed around human social relations and around parent-child relationships in particular, originally goes back to primate research. It was derived from observations of apes. Oxytocin is also responsible for social bonding in all other mammals and in birds. The same areas in the animal brain are responsible for the release of this hormone as in ours. The brain stem (that 500-million-year-old "reptilian brain") is especially involved in regulating it. Oxytocin is released, for example, when a mother or father caresses their child—even if it is an adopted child. The oxytocin is activated in the parents as well as in the child. When a person gently touches their lover or vice versa, oxytocin pours out in both partners. When a person pets an animal that they know, oxytocin is also released by both human and animal. This is evidence that there is a biological link between humans and animals.

Oxytocin is a universal bonding hormone of nature that is even active during contact between different species. I would love to know if a higher concentration of oxytocin can also be detected in the blood of someone who just hugged a tree in the woods or in their yard. That would mean that our biological toolbox of bonding activates when we encounter plants as well. I haven't come across this kind of experiment in any science journals so far, so I'm going to have to try that experiment myself.

Oxytocin receptors that react to this bonding hormone can be found throughout the entire body in humans and other animals. The immune system, muscles, heart, kidneys, pancreas, and other organs all have oxytocin receptors. Therefore, organs respond to social interaction between two humans as well as between humans and animals. That is an eco-psychosomatic mechanism. The bonding hormone oxytocin is one of the interfaces through which our social experiences with animals can affect our physical and mental health. Oxytocin translates the natural healing code of an encounter with an animal into the organic. But that's not all.

Oxytocin stimulates our good old friend, the parasympathetic nervous system of rest. Contact with animals produces the same effect as contact with trees. Reactions to stress in the nervous system of excitement are inhibited, and our body goes into relaxation, healing, and regeneration mode. That is the biological program of regeneration and new growth of healthy cells. The activity of oxytocin boosts the digestive organs and supports healing and learning processes. From behavioral biological experiments, we know that oxytocin calms fears and is effective against panic attacks. It makes our immune system more active, inhibits infections, and even increases our tolerance of pain. This explains why doctors at hospitals that offer animal-assisted therapy notice in patient charts that the need for painkillers drops when patients have contact with animals.

Oxytocin influences almost all bodily functions in a balancing, positive way. In fact, several clinical studies showed that administering oxytocin via the nasal mucous membrane enhances social skills—which isn't surprising if you consider its function as a social bonding hormone. In these experiments, participants' ability to determine another person's mental state based on their voice and facial expressions improved after receiving oxytocin. Notes on eye movement indicated that those under the influence of oxytocin paid more attention to the other person's eyes and looked longer at them.[28]

People on the autism spectrum often have low levels of oxytocin. Contact with animals increases hormone production, which is the reason animal-assisted therapy is so successful in autism. Oxytocin works especially well against depression in women with premenstrual syndrome (PMS) and encourages generous behavior as well as trust in men!

The fact that human and animal "oxytocin systems" match is another clear indication of our evolutionary and biological relatedness. I don't necessarily agree with the typical interpretation in biology that claims that these hormones "control" social behavior. It's more about substances that *accompany* our social life. Humans and animals are the ones that approach each other and actively interact with one another. Hormones also don't "command" bonding between loved ones and between parents and children; we decide for ourselves with whom we bond. Common biological mechanisms support us in doing this. Oxytocin doesn't "force" us into partnerships and relationships. We aren't machines that chemical substances control, and animals aren't machines either.

In a textbook about behavioral biology, someone seriously wrote that oxytocin bonded "brains" with offspring or sex partners. This reductionistic view reduces a human being to its brain. Modern biology is full of these kinds of interpretations that reduce living beings to mere machines, thereby breaking down every social gathering or any feeling of compassion into biochemical forces. But if you take a close look, nature is full of evidence that this biological reductionism represents one of the greatest mistakes of our time. Life is still one big mystery, even for biologists. This knowledge might be one of the most important insights that eco-psychosomatics has to offer. That is what the next chapter is about.

 ## THIS CHAPTER IN A NUTSHELL

When Charles Darwin discovered in the nineteenth century that the diversity of species had not appeared on earth after a one-time act of creation, but rather through a continuous adaptation of living beings to their environment in a millions-of-years-long process of evolution, many of his contemporaries rejected his ideas. Similarly, when today's biologists point out the close relationship between humans and animals, they often receive a reaction of skepticism or annoyance.

Based on modern evolutionary and molecular biological knowledge, *Homo sapiens* falls under the category of primates called "dry-nosed" primates (haplorhines) in biology. Our genetic code matches that of chimpanzees and bonobos by 95 to 99 percent. The way the human body is built and functions corresponds in large part to other mammals, and there are many similarities with even more distantly related vertebrates. Zoology, the biology of animals, uses humans along with other animals as subjects of study. Many scientists who conduct research on the anatomy and function of human organs at medical universities were trained as zoologists.

Humans and animals are beings of nature and come from the same network of life. Therein lies the great healing potential of encounters with animals. Because of our biological similarity, we share brain functions with many mammals and even with birds that support us in social coexistence. One of the tools for healing social encounters is oxytocin, a hormone that helps with bonding between people who are close as well as with bonding between humans and animals. That is why friendships between humans and other animals are possible.

Contact with animals—as we already know about contact with trees—leads to many positive effects on our health that can be used in medicine. When we interact socially with animals and pet them, it measurably strengthens our immune system. Animal-assisted therapy increases the chance of surviving a heart attack and is also implemented in treating persistent vegetative state (PVS) patients. Numerous medical effects on body and mind from contact with animals have been verified. One reason for the healing effect of animals is oxytocin, which is released when we interact with them. We have oxytocin receptors throughout our body and in our organs that react to the hormone in our blood. This is a mechanism though which social contact with animals measurably affects our physical health. Clinical studies have shown many healthy effects of oxytocin on our heart and mind, for example, but it is only one piece of the puzzle in explaining animal-assisted therapy. Like a forest, animals also have a holistic effect on us. They activate our biophilia forces.

Our relatedness and social connection with animals should never be judged as an insult to human pride. Instead, it should be an opportunity to see animals as worthy counterparts.

# 6

# THE BIG SECRET OF LIFE

## Life Forces

Recall Jakob Johann von Uexküll, the Estonian zoologist and philosopher who in the nineteenth century introduced the idea of the functional circle between living creatures and their environment and said that our body has no fixed external borders. The assumption that evolution advanced not merely by coincidence but, above all, by the active reaction of living beings to natural influences, is now supported by the field of epigenetics. Von Uexküll argued that biology is fundamentally different from physical sciences, such as physics, chemistry, and mineralogy, all of which investigate inanimate matter. In all living creatures, from unicellular organisms to *Homo sapiens*, a special life force or principle of life has its effect.

Jakob von Uexküll and other proponents of this idea were criticized by other scientists of their time. They were called "vitalists." This sounds harmless, but it was a derogatory term in the nineteenth century. Von Uexküll, whose life spanned from the nineteenth to the twentieth century, acknowledged the theory of evolution founded by Darwin in the nineteenth century. He did not dispute the fact of evolution, but merely challenged a strictly materialistic view of the phenomenon of life.

Joachim Bauer, a German molecular biologist, medical doctor, and professor at the University of Freiburg, is also a top researcher at the Mount Sinai Health System in New York. He sees in evolution a *creative process* of living beings themselves.[1] James Shapiro, professor of genetics and microbiology at the University of Chicago, also concluded from the latest molecular biology findings that living beings control the evolutionary process in a creative way.[2] This position contradicts the majority of biologists' conception of life because it suggests something like a nonmaterial life force.

The Greek philosopher Aristotle (384–322 BC) believed in a principle of life that is not material. According to Aristotle, all forms of life, whether human, animal, or plant, have a "soul," and the body, together with the soul, forms an inseparable unit. Modern psychosomatics now provides evidence for the inseparability between soul (from the Greek "psyche") and body ("soma"). Aristotle did not use the concept of soul in the religious context that it connotes today, but as a principle of life that is at work within us and that is at the same time also the *cause* of life. He wrote, "Therefore the body as well exists somehow for the sake of the soul, and the parts are for the sake of the functions towards which each of them naturally grows."[3]

Biology of the third millennium is increasingly oriented toward reductionism. This is a scientific approach in which complex phenomena such as life are reduced to their individual parts. Chemical and physical processes are then not seen as concomitants of life; instead, some think that life *is* nothing more than these processes. According to this idea, love and bonding between two people or between a human being and an animal is merely the biochemical activity of the bonding hormone oxytocin and other substances. This is how statements by behavioral biologists appear in textbooks claiming that the bare chemistry of oxytocin bonded our "brains" together.[4] One of the greatest mistakes of modern biology is the dogma that we *are* our brain. This dogma results from the approach of reductionism. Breaking down research objects into their individual parts in order to better understand them is quite useful; however, where reductionism goes wrong is when we forget to switch back to the higher level of holism. In the course of this chapter, I will address this problem in more detail.

The distinction between the Aristotelian concept of soul and religious content is important, no question, because the great philosopher was also a naturalist and did not have religion in mind with his ideas of the soul. His philosophy, however, underlines the fact that there is an untold secret of life. He assumed that there was an "unmoved mover." The unmoved mover is the cause of the world and of life, but has itself no cause and cannot be moved by any other cause. The unmoved mover brought *eidos* into the world; according to Aristotle, this is the formative power that acts as the opposite of matter in living beings. Eidos is, therefore, not material.

Modern biologists and biophilosophers justifiably distance themselves from looking for religious proof in the philosophy of Aristotle. Many of them, on the other hand, are far less cautious when they want to use the ancient Greek philosopher to defend their assumptions. The ventured hypothesis that Aristotle had DNA in mind when he spoke of the formative power of eidos is making its rounds among biologists. DNA (deoxyribonucleic acid) is the molecule that carries our genetic information, that is, the material substance of our genes. Some biologists, such as the geneticist and Nobel laureate Max Delbrück, even compare DNA with the role of the unmoved mover because our DNA transmits a plan for the movement of proteins from which the living body grows, while the DNA itself—in its chemical basic structure—does not change; in other words, it is not "moved."[5]

Delbrück, who died in 1981, conducted research at a time when biologists thought they could find the last answers to the great mysteries of life in DNA and thus decipher the phenomenon of life. The notion that a chemical molecule can embody the principle of life, eidos or the unmoved mover from the Aristotelian philosophy, seems to be a daring one, even for the time. DNA itself is matter and had to develop over the course of evolution. It cannot be the cause of life. Besides, Aristotle had something *intangible* in mind about the principle of life, thus not a chemical molecule.

In the meantime, several decades have passed in biology laboratories. The bold predictions of former geneticists that we would soon be able to decipher the human genome have not come true. This does not prevent biologists such as the world-famous Oxford professor Richard Dawkins from declaring genes to be the only relevant life force at all. In this book, I have mentioned arguments that suggest evolution is advanced by the active reaction of living creatures to their environment—because they participate in a functional circle with nature. Genes would, therefore, be tools of evolution. Species would be the real actors to which biologists such as Joachim Bauer and James Shapiro give credit for the creative contribution of their development. Dawkins and other "ultra-Darwinists" see it exactly the other way around. They say that evolution proceeds only on the level of genes. Dawkins introduced the concept of the "selfish gene." According to this idea, life and evolution serve selfish genes to multiply and reproduce. Suddenly, DNA is not a tool for the species to evolve, but rather the species are declared to be tools of DNA molecules. Their sole purpose is to enable DNA reproduction.

This extreme reductionist view of the phenomenon of life is called "gene-centrism." Evolution is, from that view, exclusively about genes, while species are only a byproduct of the evolution of genes. Although criticized by many biologists, the representatives of this school of thought have succeeded in creating a large media following. Dawkins, for example, was described by the German news magazine *Der Spiegel* as "the most influential biologist of his time."[6]

As the above examples show, the idea that DNA is the true life force—the essence of living—is encountered in various ways in biology. When biologists try to interpret and explain life on a purely material level, it sounds like something out of a wild science fiction fantasy world. Did you know that some biochemists believe that the first life-form on earth was a kind of tiny water puddle on a stone or came out of a piece of clay? Follow me on a journey to the beginnings of our universe and then once again to the time of LUCA, the first of all living beings.

## The Mystery of the Origin of the Universe

For many generations, people have asked themselves why there is a universe at all and not a gaping nothingness. In the seventeenth century, the German philosopher Gottfried Wilhelm Leibniz (1646–1716) asked the question: "Why is there something and not nothing?" Another German philosopher, Friedrich Schelling (1775–1854), then asked something similar in the eighteenth century: "Why not nothing? Why anything at all?" In the twentieth century, the philosopher Martin Heidegger (1889–1976) joined the others with his question: "Why is there existence at all, and not nonexistence?"

The physics professor Lawrence Krauss at Arizona State University in Phoenix finally enlightened the world about the origin of the universe in 2012 and cleared up the largest question of all humanity. This is remarkable since, due to physical and mathematical reasons, no one has access to data from the very first split seconds after the Big Bang, an estimated fourteen billion years ago. For this reason, virtually all cosmologists believe that we will never be able to reconstruct the first of all moments—if there really was such a "first" moment at all. Apart from this, our research can never refer to a "before" or "beyond" for justifiable reasons. The physical laws of the universe do not exist before or after. They were produced together with matter, however this may have happened. For this reason, mathematics cannot allow a view behind the curtain, either.

All of that didn't keep Professor Krauss, who along with Richard Dawkins forms a kind of spearhead of radical materialism—the doctrine that physical matter is the only reality—from giving his now world-famous book the title *A Universe from Nothing*. I read it. And it turned out the title is false advertising. Krauss does not explain how a universe could emerge out of nothing, but merely places his argument on the table, which is that, in his view, the formation of a universe can be triggered by very few preconditions. This is a striking difference. The book should therefore actually be called "A Universe from Just a Little."

"Little," is not "nothing," and the question of the origin of our world remains unanswered. The "nothing" of Lawrence Krauss's book consists, after all, of a plethora of quanta, which are tiny energy packages that can appear and disappear. This means that to form a universe from this "nothing," which in reality is a "something," a kind of matter and physical energy must already have existed. The question also remains as to where these "nothing-quanta" were supposed to have been, for even a vacuum containing the quanta would have to be somewhere. If one says that "only" quantum energy is needed to create a universe, one must still explain the occurrence of these mysterious quanta. David Albert, a renowned professor of theoretical physics at Columbia University in New York, tore apart the theories of Lawrence Krauss about the "universe out of nothing" and called them untenable. This statement appeared, among other places, on March 23, 2012, in the *New York Times* under the title "On the Origin of Everything."[7]

That there is matter at all in our universe is due to another cosmological mystery. When the universe was one-millionth of a second old after the Big Bang, the entire resulting matter was—according to current thinking in astrononmy—compressed into an extremely dense plasma that, according to astrophysicists' calculations, reached an inconceivably hot temperature of 10 trillion degrees Celsius. That is a one with thirteen zeros. All the planets, moons, stars, meteors, and creatures of our universe—you and I—are now made of this plasma.

The particles swirling around began to annihilate each other immediately after their emergence. There was an antiparticle for almost every particle of matter, and when the two crashed together, they mutually eliminated each other. Scientists are faced with the great mystery of why, after the large obliteration of particles, about one-billionth of the original matter remained. If it hadn't been for that additional matter, our universe would be empty and only an echo of the Big Bang. This peculiar imbalance resulted in the fact that we exist at all. There was just enough matter left necessary for a stable universe suitable for life—not too much and not too little.

Whenever something happens that is highly improbable, yet absolutely necessary for the origin of life, scientists have to revert back to the anthropic principle in the absence of any other explanation. As a reminder, anyone who uses the anthropic principle puts the cart before the horse to avoid asking for explanations for things that *cannot* be scientifically explained at the moment. They will say, "If it had not been so, we would not be alive to ask this question." We are here and can observe the universe simply because it has qualities that make it possible for us to be here and observe the universe. If it were not for these qualities, there would be no observers either. As we can clearly see, the anthropic principle does not provide explanations for the causes of unlikely events. It simply accepts them as a given.

In order for life to occur, countless events were necessary over the course of fourteen billion years, each one being extremely unlikely. I call these events "bottlenecks." Even before *Homo sapiens*, the world had to overcome a nearly endless obstacle of the bottlenecks that we currently can only explain with the anthropic principle. A well-known bottleneck is the question of fine-tuning laws of nature. The slightest deviation in any physical law of nature, such as gravitation, would have made it impossible for a stable universe to form that also supports life. The anthropic principle is once again satisfied with saying that if the constants of nature were not exactly aligned as they are, we would not be here to deal with the question of fine-tuning the constants of nature.

Because these kinds of "answers" are not satisfactory for most scientists, some physicists look for ways out of this dilemma. The physics professor Max Tegmark at the Massachusetts Institute of Technology (MIT) theorizes that there are an infinite number of universes. This is called the "many-worlds interpretation" or the "multiverse theory." The professor's trick is claiming that an infinite number of constellations of laws of nature exist in an infinite number of universes. The likelihood that the constants of nature are as perfectly aligned with one another as they are here is still very low for every single universe. If, however, there are infinitely many universes with infinitely many combinations of attributes, then universes like our own that can sustain life must be among them.

The many-worlds interpretation, for which there is not the slightest evidence, is pure science fiction. There are even some physicists who posit that our universe is actually an experiment conducted by intelligent creatures from another universe.

Neither the many-worlds interpretation nor the belief that we are an experiment of aliens brings science further in the investigation of causes. On the contrary, we are dumping the question of origin into a dimension outside of our universe, where it would no longer be accessible to us.

Particularly mysterious is a phenomenon that took place in the early stages of our universe. Astrophysicists assume that sound waves dispersed throughout the universe were similar vibrations to those of a musical instrument. This "music of the early universe," with its tones and overtones, contributed to an organizing and structuring process of matter, which led to the formation of areas with higher and lower density. Only then could galaxies and solar systems form.

Since bodies of humans, animals, plants, and all other known forms of life are based on carbon, the universe had to develop a sufficient amount of this chemical element so life could develop. This is another bottleneck. Stars produce the carbon on which our life is based. Hence, we are all made of "star dust." Carbon as the foundation of life could only be formed in sufficient quantities because in large stars a complex nuclear fusion occurs, which is possible because of certain properties of helium. If these helium properties were only slightly different, there would not be enough carbon available, and our existence would be impossible. This too is a case for the anthropic principle. Over the course of creation of the universe, one precise event after another took place, all of which were necessary to enable life. In the third millennium, our universe is still a place of undiscovered secrets. One of the greatest mysteries is and remains life itself.

## The Spark of Life

Let us leave behind the many bottlenecks the cosmos had to go through during its development in order to become a universe that inherently allows the evolution of living beings. Let us return to life, which is the focus of this book. Unlike what the zoologist Jakob von Uexküll postulated, biology looks at life as a purely material phenomenon, so scientists are searching for explanations to answer the greatest question of biology: How could life-forms develop from inanimate matter? The birth of LUCA, the earliest ancestor of all creatures, is and will likely always remain an unanswered mystery.

In order to consistently rule out any consideration of a nonmaterial life force from the beginning, biologists take a stab in the dark, fantasizing like Lawrence Krauss with his "nothing-quanta," from which the universe supposedly emerged. One hypothesis that sounds relatively plausible claims that the first living creatures on Earth were transported to our planet on meteorites from the depths of the universe. It is possible that bacterial spores or single-celled organisms were brought to the earth in this way. But even if this is true, it is by no means an explanation for the *origin of life*. The question of what the first creatures emerged from is not answered, but merely transported into another realm of the universe.

The "primordial soup theory" also sounds somewhat credible. It was introduced into the world of science by the American biologists Stanley Miller (1930–2007) and Harold Urey (1893–1981) in the 1950s. It asserts that the first organic molecules were created four billion years ago by chemical reactions out of water, ammonia, hydrogen, and methane under conditions of the primordial atmosphere. Lightning supposedly gave water so much energy that chemical reactions occurred that otherwise wouldn't be possible. This chemical evolution in primordial waters of the earth is said to have led to the entire biodiversity of our planet.

Biologists simulated the conditions of primeval Earth in experiments and found that, in fact, very simple biochemical compounds could have been produced that also occur in living cells. As early as in the 1950s, editors of the *New York Times* announced that biology was already on the verge of explaining the origin of life. Six decades later, we are not a step further. No experiment has ever succeeded in producing even the simplest form of life from chemical reactions.

Hypotheses of biologists in the 1970s and 1980s were especially quirky. Günter Wächtershäuser, a visiting lecturer for evolutionary biochemistry at the University of Regensburg in Germany, theorized that the first organism on earth—that is, LUCA—developed from nothing more than a kind of water puddle on the surface of a stone on which a "surface metabolism" had occurred. These were supposedly simple chemical reactions from which later the metabolism of the first cells arose because the surface of the stone gave the chemical processes a physical frame.

This explanatory model also presupposes that—to put it crudely—a puddle eventually turns into a primitive living being, as long as chemical reactions keep occurring. There is no evidence that such a thing is possible. Besides that, these "metabolism puddles" cannot reproduce, and this is an indispensable prerequisite for life. And this "stone puddle world" could never explain the origin of DNA.

Another attempt to reduce the formation of the first living cells to the most trivial material processes comes from Doron Lancet, a professor of biochemistry at the Weizmann Institute of Science in Rehovot, Israel, and biochemistry professor David Deamer at the University of California, Santa Cruz. The two proposed the lipid-world hypothesis. Lipids are fatty substances that have a water-loving (hydrophilic) and a water-repellent (hydrophobic) end. Think of the oil in your salad dressing, which forms beads. This occurs because of the lipids in the oil. The beads are formed because the water-repelling ends of the lipids turn inward, to move away from the water, while the water-loving ends are directed outward toward the water. This form, derived from the Latin *mica* (meaning "grain" or "crumb") is called a "micelle." The external borders of living cells also contain lipids, similar to oil beads in salad dressing. Therefore, Lancet and Deamer consider it possible that the original cell LUCA emerged from an oil bead in the primordial soup, in which chemical substances were enclosed.

According to the lipid-world hypothesis, a bead of oil in the primordial ocean four billion years ago might represent our oldest ancestor. Even if one assumes that a large number of molecules were coincidentally enclosed in this LUCA-bead and could temporarily form a kind of metabolism, how it could then turn into a living cell capable of reproduction and with its own DNA remains a complete mystery—a cell that was able to evolve, and billions of years later, even become human. The chemistry of oil beads in the primordial soup does not have the slightest resemblance to the chemistry of cells, and certainly not to the chemistry of heredity. LUCA was definitely not a primeval stone puddle or a floating bead of oil.

In the 1980s, the idea of a primeval "clay being" spread among biologists. According to this hypothesis, LUCA had no metabolism at all, even though the existence of a metabolism has been defined in biology as an essential criterion of living creatures. The chemist and molecular biologist Alexander Graham Cairns-Smith at the University of Glasgow suggested in 1985 that the forerunner of all forms of life were clay minerals. Clay is a combination of mineral materials with different particle sizes, such as sand and silt, which were present in great quantities in primeval times. Their chemical structure is a crystal, and crystals can grow by absorbing water and mineral particles from the environment and incorporating them into their structure. That is, they are able to pass on their chemical structure to their mineral "descendants." Defects and changes in the structure of the clay mineral can also be passed on to a certain extent. In this, Cairns-Smith sees parallels with today's genes, which also pass down their blueprints. The clay minerals supposedly formed a kind of framework for later life. Because they can spread their structures, they are—according to this bold hypothesis—capable of a simple form of heredity and evolution.

It is highly unlikely that this claim is true and that LUCA or its preliminary phase was really a clay being. Clay minerals are inanimate substances; they are inorganic molecules, not organic ones. The notion that living beings could have emerged from clay particles is a downright negative spin-off of the doctrine of materialism that has spread in biology. Obviously, no experiment has ever succeeded in making living organisms from clay minerals. Clay minerals would also not be able to "inherit" their crystal structures with sufficient accuracy to initiate an evolutionary process. In addition, it is completely incomprehensible how a clay mineral could end up having a metabolism, forming cells, or producing DNA. There is no trace of a "clay metabolism" in today's living beings.

Francis Crick (1916–2004), a physicist and biochemist at the University of California, San Diego, assumed in the 1960s that the first living being was a simple gene. Crick was one of the discoverers of DNA, which is why it is not surprising that he focused his attention entirely on genes. In the 1980s, other biologists then took the idea that LUCA could have been a single gene and spoke of the "RNA world."[8] But how did this "living" primeval gene supposedly appear?

In the primordial soup four million years ago, simple molecules apparently emerged, which were precursors of today's DNA. At least two of these molecules would have accidentally united, and around them a simple membrane (that is, an outer cover) would have formed. How all of that could have happened, nobody can say. Later, the molecules and their covers supposedly developed into complex cells with their own metabolism and their own DNA.

No experiment has ever produced this kind of living gene. In addition, it is extremely unlikely that such gene-like molecules—which could successfully reproduce themselves—formed accidentally in the primordial soup. If the emergence of life had been a purely material process, only chance could be a trigger for life processes. The biochemist Kevin Plaxco at the University of California, Santa Barbara, wrote a book on astrobiology, which is the exploration of the emergence of life in the universe. He describes how in the 1980s he met Harvard professor and Nobel laureate Walter Gilbert, who had coined the term "RNA world." The professor told Plaxco about an experiment he was working on at that time in his laboratory. He wanted to see whether he was able to make a kind of "living gene" through chemical reactions in a test tube. That was thirty years ago, and so far this experiment has yielded no results. The idea that our ancestor LUCA was a bare gene molecule in the primordial soup remains wild speculation, just like all other attempts to derive the initial ignition of life from nothing but matter. Life remains an undiscovered mystery.

The physician, philosopher, and professor of psychiatry and psychotherapy Thomas Fuchs at the German University of Heidelberg counters this obstinate reduction of the living to mere matter. He writes, "Physical descriptions and explanations should then apply to all areas of life. The green tree is only a large cluster of molecules, the song of the nightingale in its branches is a merely irregular sequence of air pressure fluctuations, and the joy of the wanderer who listens to it, a certain neuronal pattern of excitement."[9]

## Why Human Consciousness Is Connected to Nature

Immanuel Kant (1724–1804) rebelled against the notion that living creatures can be explained purely materially. He saw in living beings a "formative power" that matter does not have. He described it as a "self-propagating formative power, which cannot be explained through the mechanism alone."[10]

One physicist who was especially enthusiastic about the mystery of life was Albert Einstein (1879–1955). He wrote that the scientist's "religious feeling takes the form of a rapturous amazement at the harmony of natural law, which reveals an intelligence of such superiority that, in comparison with it, all the systematic thinking and acting of human beings is an utterly insignificant reflection."[11]

Einstein spoke of a form of religious feelings that the cosmos and life triggered in him: "I maintain that cosmic religiousness is the strongest and noblest incitement to scientific research. . . . A contemporary has said, not unjustly, that in this materialistic age of ours the serious scientific workers are the only profoundly religious people."

Let us add another piece of the puzzle from the original German texts of Einstein in order to get a picture of the kind of spirituality the influential physicist had in mind. "How can cosmic religious feeling be communicated from one person to another, if it can give rise to no definite notion of a God and no theology?" he asked in his essay "Religion and Science."

Einstein used the term "cosmic religiosity," but we could also refer to it as "cosmic spirituality" or "spirituality of nature." He made it clear in the same breath that he had no theology or notion of God in mind. He did not speak of a confessional religion. Einstein was a pantheist. He argued that God or a creating energy was one with the cosmos and one with nature. From a pantheistic point of view, an encounter with nature is an encounter with God. Pantheists like Einstein have no personified "being" in mind when they use the term "God." On the contrary, the divine exists in all things. It is the spiritual source of the cosmos and life.

Today, six decades after his death, few scientists represent a pantheistic point of view. Einstein was anxious to maintain "cosmic

149

religiosity," when he wrote, "It appears to me that the most important function of art and science is to awaken this feeling among the receptive and keep it alive."

Ernst Pöppel, a professor of medical psychology at the Ludwig-Maximilian University of Munich, wrote in 2011 about the origin of life: "How this could happen, or what the driving force behind this was, remains the biggest mystery to me. How was it possible that molecular compounds created something to which we attribute the quality of life?" These lines were written by Pöppel in his essay titled "We Are Meant."[12] The essay appeared in a scientific book on spirituality. Nature, the origin of life, and evolution of species—these are the themes that represent a spiritual dimension for most people. Atheists also describe spiritual sensations that nature triggers in us. Even the philosopher Michael Schmidt-Salomon, who is referred to in the media as "Germany's chief atheist," reported on a spiritual experience he witnessed through nature. He described his sensation as an "oceanic feeling."[13]

In his essay, Pöppel also links human spirituality with nature and evolution of life. "Single-celled organisms entered the world and were extremely successful. If we ask ourselves how the simplest life-forms are organized, then we find something remarkable: from the very beginning of life, organisms are characterized by what they do, as humans are. If you want to be exact, very little has come along over the course of evolution. What distinguishes us in our experiences is billions of years old."

Now we are, once again, very close to the healing code of nature. Life processes in nature have a lot to do with us humans. They created us. We recognize nature as our home and other forms of life as our relatives. Our identity is nature—not concrete and asphalt. Above all, we are connected with nature through our consciousness because nature produced the human consciousness. In nature, we can learn more about our consciousness than in any other place. Even the simplest single-celled organisms react to their environment. They recognize different wavelengths of light and know where to go when something to eat is close by. They look for sex partners in their water world for breeding.

The interaction of unicellular organisms with the environment *is* a very simple form of consciousness. The complex consciousness of humans and animals developed over time from the primordial consciousness of unicellular organisms. Our consciousness can be traced back to LUCA's initial moments of primitive consciousness.

Let's take a closer look at this process. The descendants of LUCA (that is, single-celled organisms) joined together to form colonies of cells billions of years ago. Spherical algae, which biologists call volvox, is a transitional form between single- and multiple-celled organisms. They still exist today. Volvox is a sphere made of a few to thousands of algae cells that have joined together to form an organism. Together they become a sphere with a diameter of one-tenth of a millimeter to one whole millimeter. Most of the volvox cell colonies are only visible under a microscope, but their largest representatives are recognizable as tiny dots in the water.

Each individual cell of the volvox has its own eye spot, with which it can distinguish between light and dark. And each cell has two organs of movement that look like whips, organs for photosynthesis, and a very simple organ for the excretion of water if too much has entered the cell. The water is absorbed by a tiny bladder, which contracts to press the liquid outward again. Each of the cells from the colony could survive on its own. They are equipped like conventional single-celled green algae, which innately live on their own. Sometimes the volvox bursts. Then the individual cells simply continue to live as green algae. But united in the volvox, they divide up different tasks among themselves in order to keep their common sphere alive. In the front area, cells provide for the movement of the volvox, which moves through the water with rotary motions. This target-oriented swimming is possible because all the cells beat their whips together synchronously. They communicate with each other via "plasma bridges" (threads of cytoplasm that pass through the walls of adjacent cells) and mutually coordinate their swimming movements.

I remember when I saw my first volvox under a microscope at the beginning of my biology studies. I was fascinated by the sight of the bright green ball. It was fast, and I had my hands full to keep it under the lens. I could easily see individual cells, and the volvox moved gracefully through the water.

The cells jointly perform the task of forming the extracellular matrix. This is the substance inside the volvox, in which the cells are embedded and which holds them together. In other words, they provide whatever is necessary for their connection by themselves. They actively stick together. The cells also carry out photosynthesis together. That means they absorb sunlight and use this energy from water and carbon dioxide to produce high-energy carbohydrates, just as all green plants can. In another area of the volvox, the cells are free from all these tasks. They dedicate themselves to a combined effort of reproduction and ensure that their mutual way of life is passed on. They divide and form a small volvox embryo. They connect the cells of the embryo via plasma bridges. The mother ball thus produces a finished new ball of single-celled organisms. The small cells come into the world already in the volvox: the mother ball releases its sprouts to live their own life. And if the volvox ever bursts, each individual cell can live as an independent green alga, just as the cells in the parent volvox can.

Whether the volvox is regarded as one organism or as a colony of single, closely cooperating green algae is a matter of opinion. It is both at the same time. Billions of years ago, similar collections of cells like the volvox represented a milestone in the development of life. Why did they emerge? It was a strategy of the single-celled organisms to better survive in their primeval water world and to give themselves an advantage. It is easier to produce energy, find food, move around, and reproduce as a collection of cells.

The phenomenon of cell colonies is astonishing because it shows that living organisms actively prodded evolution at an early stage and didn't just wait for chance mutations. In order to live as a volvox or any other cell colony, it was not enough that cells were simply washed together by currents in the water. The individual cells had to react, work together, and maintain their connection. As we have seen, they even had to make sure their descendants practiced the same way of living together as they did themselves. They had to pass on the *traditions* of the volvox. All of these abilities were interwoven with the environment in which the respective cells had to survive.

Multicellular organisms were formed from cell colonies similar to the volvox. Later volvox forms became real multicellular organisms. The cells split the work among each other more and more until they were so specialized that they could not survive on their own. At this point they were completely dependent on each other. They had become one. The disadvantage to this was that it meant death for a cell if it left the organism. This disadvantage was, however, offset by one serious advantage. Once the cells specialized, they could turn to increasingly complex tasks. Microscopically small multicellular organisms formed at first. In the course of evolution, whole organs and organ systems emerged from the specialized cells. Fundamental functions of living beings were already inherent in single-celled organisms and cell colonies such as the volvox. They existed as a kind of idea or precursor to later organs. The fact that today's mammals no longer have an eye spot, but complex eyes, is due to the progressive division of the cells' work.

Today our cells are still connected to each other via plasma bridges and exchange information via these connections. Neighboring body cells communicate with each other, just as the cells of the volvox do. Our cells are also embedded in an extracellular matrix, which gives them a common form. It is like early cell colonies in the water, but everything is much more complex. Like volvox, we also have special cells that are solely responsible for reproduction. They produce egg and sperm cells; for their formation, a kind of cell division is necessary, which is completely different from the division of body cells. We exist because cells have actively remained together and collaborated for over four billion years. Throughout the evolutionary process, cells worked together in interdependency with the environment and actively adapted to environmental conditions. From the first single-celled organism on, it was necessary to perceive environmental stimuli, so cells could react to them. This was what Jakob Johann von Uexküll meant when he spoke of living beings' "perceptive organs" in a functional circle with their environment.

The perception of the environment was thus a whole-body process from the beginning of life. Single-celled organisms did not have a brain yet. They perceived the stimuli of nature in a very direct, primitive way, and responded to them. Single-celled organisms were certainly not capable of mental experiences, but they were directly *aware* of their surroundings, and this is a very simple form of consciousness or at least a precursor to it. Our own consciousness can be traced back to single-celled consciousness. It has always been connected to nature.

This realization has far-reaching consequences for the self-image of *Homo sapiens* at the beginning of the third millennium, which is dominated by materialistic perspectives on the phenomenon of "life." In the computer age in which we live, many scientists believe that we are "biocomputers" and that our mind is a kind of software that runs in the brain. But from nature and its life processes, we can learn a different perspective, as we shall soon see.

## Nature Shows Us We Are Not the Brain

The fact that we are capable of more conscious activities than single-celled organisms is related to the division of labor between the cells, which mutually created increasingly more complex organs and functional networks. This also increased the abilities of life-forms to interact with the environment and other living beings. Single-celled organisms sensed contact with the outside world through simple chemical reactions on their outer shell. Over the course of evolution, nerve cells arose to intensify the perception of the environment. They spread like sensors across the entire body. New sensory stimuli arose, and the ability to feel was born. But there was no center for a long time where the stimuli could be processed. The nerve cells continued to develop. Evolution "invented" ganglia. These are nodes in which the nerves are particularly closely intertwined. The ganglia spread over the entire body. Earthworms, for example, contain ganglia in each of their body segments, which are interconnected by nerve fibers. Our human nervous system also has ganglia at different points in the body. They are control centers, and biologists consider them to be precursors of the brain.

The organization of these control centers continued to progress. The first brains emerged from large ganglia. Our mammalian brain is ultimately a highly compressed and unimaginably complex differentiated "node" of the nervous system that is assigned the task of control center, just like the ganglia in earthworms.

The degree of consciousness of living beings increased continually in this way over the course of evolution. There is no turning point in the history of life that marks a boundary between "not conscious" and "conscious." Consciousness was in life-forms from the very beginning, even in single-celled organisms, which set the gradual development of the consciousness in motion through their interactions with the environment. Of course, this was a very elementary consciousness, a "consciousness seed." Monkeys, on the other hand, are almost as conscious as humans, and other mammals also have high levels of consciousness.

Here is a quick summary: The emergence of consciousness can be traced back to single-celled organisms and resulted from the interaction of living beings with their environment. Because the cells and organs of multicellular organisms became increasingly specialized, highly complex nervous systems formed, which condensed, leading to the creation of nodes and eventually the brain.

Our brain is therefore an organ of interaction with the environment. Above all, it is embedded in the complexity of the whole organism. Consciousness always includes the whole living being, never the brain alone. Contrary to the predictions of some neuroscientists, we will never find an area in the brain in which consciousness is "executed." Consciousness is not generated by an organ; it is created when a holistic organism interacts with its environment. The brain is an organ of relation. It allows us to relate to our habitat and other living beings. Other organs and systems of the organism are also involved in the development of consciousness. A brain on its own is neither conscious nor able to enter into relationships, nor even formulate a thought for that matter.

The writer Diane Ackerman teaches at Columbia University in New York. She has made a name for herself by translating scientific knowledge into easily comprehensible language. She also asked if our consciousness was seated in the brain and came to the following conclusion: "Most people think of the mind as being located in the head, but the latest findings in physiology suggest that *the mind* doesn't really dwell in the brain but travels the whole body on caravans of hormone and enzyme, busily making sense of the compound wonders we catalogue as touch, taste, smell, hearing, vision."[14]

We also know today that we have a dense nerve network in the abdomen that performs like a brain. That is why biologists speak of the "abdominal brain." The abdominal brain is another independent center of the human nervous system. Recent research suggests that it is involved in the development of emotions and moods. The phrase "gut feeling" is therefore not far from the truth. Even mental disorders can be caused by the abdominal brain. Biologists call the functional connection between the brain in the skull and the brain in the belly the "gut-brain axis."

Communication between the abdominal brain and the brain occurs not only via nerve cells and hormones from the gut but also by means of signals from intestinal bacteria. The microorganisms that colonize our intestines thus become part of our bodily functions in several respects. They support the immune system, help with digestion, and participate in the communication between our abdominal brain and the "head brain."

The physician, anatomist, and physiologist Hans Jürgen Scheurle describes the interaction between the environment, the brain, and the rest of the body. He also repudiates that the brain "produces" our consciousness, and sees the role of the brain more in awakening the potential and abilities of the entire living being: "Thus, fear and joy are not based in the amygdala (in the brain) but in the whole body, in the eyes, and also in the heart, as love songs correctly claim. They are only awakened by the brain nuclei."[15] Scheurle, who has also written several medical textbooks, calls the assertion that nature and living beings can be reduced to matter alone an "annoying paradigm that, just like religious dogma, does not like to be questioned."

Sensing, perceiving, feeling the environment, social interaction—all these are experiences of the consciousness of the *entire* body and not solely the brain. Molecules and chemical chain reactions alone are not capable of these experiences. By interacting with other people and animals as well as with our environment, we actively maintain our awareness. As I said, the brain is an important tool, no doubt about it, but it does not generate our consciousness and is not its home.

We cannot "watch the brain think" via modern imaging methods, as we are sometimes led to believe. The colorful images of brains presented to us in books and on television are not images of a thinking brain. These are merely statistical graphics prepared as diagrams. The colors indicate which brain areas are supplied with blood when the subjects have certain thoughts or other consciousness content. However, crude statistical blood flow patterns are not evidence that thoughts or consciousness are "generated" in these locations. They only signify that activity of the area Y in the brain is involved in consciousness content X and not that neurons there produce a thought or a certain consciousness content in the form of a chemical process. Scientists will not be able to keep their promises to understand our consciousness in the near future, to physically "find" it in the brain, to explain the origin of life from purely material processes, or read the DNA of human beings "like a book." And they've been promising the last two for decades.

The previously mentioned medical psychology professor Ernst Pöppel summed up this topic as follows: "Humans and single-celled organisms are not as different as we may think. The principle of life—to move to where living conditions are better—unites us. In order to make the comparison between 'here' and 'there,' people and unicellular organisms need organs to perceive the world. They need rating systems and categories. Our consciousness's present connection to this world results from this ancient legacy, which has been passed down since the beginning of life."[16]

## Eliminating the Mind

The German physicist Walter Heitler (1904–1981), a professor at the University of Zurich and the Dublin Institute for Advanced Studies and winner of several prestigious science awards, wrote, "The expression 'spirit' may not be popular in today's world, where an excessive doctrine of materialism is putting forth somewhat evil buds. But that's exactly why we have to understand what is natural law and natural knowledge. Nature follows the law from this *nonmaterial spiritual element*. Consequently, spiritual elements are also anchored in nature itself."[17] Heitler meant by this that the cosmos and nature can be mathematically described. A mathematical law is something nonmaterial, argued Heitler, which is evident from the fact that the human mind can recognize it. Consequently, the spirit must be at work in nature and in the cosmos.

More than forty years have passed since his statement. The term "spirit" seems to be completely out of fashion now. It has become difficult for humans to feel like a spiritual being, which—as natural beings—we automatically are. The doctrine of materialism's "evil buds" that Heitler had complained about in 1970 have grown so much in the meantime that it almost inevitably triggers existential crises in many people.

In 2013, I was invited to a panel discussion in Mozart's birthplace of Salzburg, Austria. A well-known news reporter and foreign correspondent who had reported on numerous crises and wars around the world also attended. While we were chatting after the event, we somehow ended up talking about neuroscience. My pleasant conversation partner confided that he finds it offensive when some neurobiologists assert that he, as a human being, is nothing more than the sum of chemical reactions in his brain—that he himself is practically an illusion that is created by a computer in his head. "We are reduced to machines, which I find really terrible," he said.

Hans-Jürgen Scheurle sees this as an offense as well. He expresses himself clearly against the reduction of human consciousness to brain chemistry. "The notion that a person is a subject trapped in his inner world is not a neutral claim, but rather represents a devaluation of the human mind. The theory of brain-controlled consciousness makes the individual a 'prisoner of the brain' without any real existence. It sees in the human ego an ego separated from the world instead of a universally living spirit, and grants humans only an illusionary, unreal consciousness."[18]

In their degradation of the human mind, some neuroscientists exhibit a great deal of overconfidence. For example, the well-known German brain researcher Gerhard Roth and some of his colleagues compared their findings with the achievements of Copernicus, Charles Darwin, and Sigmund Freud.[19] Copernicus delivered the first great "offense" to humankind when he realized that the earth is not the center of the universe. Later, humanity was offended once again by Darwin's realization that we are from the animal kingdom and not the crown of creation. As I've said, many people today still feel offended by this fact. The third great slap in the face to mankind came from Freud. We are not masters in our own house, but our actions are strongly influenced by our unconscious. Now, according to Roth, the next great offense of humans is brought about by the knowledge of modern neurobiology, according to which the ego is a pure construct of the brain—that is, an illusion based on chemisty. There is nothing spiritual, only matter. There is no real agent inside us, no protagonist. Free will, too, is an illusion that our brain pretends we have.

Over forty years ago, when the physicist Walter Heitler bemoaned "excessive materialism" and its "somewhat evil buds," he was not yet aware of the buds the doctrine of materialism would still be putting forth at the beginning of the third millennium. One of the most objectionable of these buds is certainly the reduction of human and animal minds to pure brain chemistry, as I just described. Einstein's dream that science would pass on the "cosmic religiosity" from person to person has apparently not yet been fulfilled.

I dare to assert that our upbringing and education system planted this worldview of materialism deeply within us. The latest dogma—our mind is software installed on a biocomputer, and everything living is essentially nothing but matter—affects the human self-image and the way we perceive nature. Nothing truly living is left. The abolishment of the mind seems to have been successful.

I once attended a conference of psychotherapists as a guest speaker. Some reported how the idea of being only a brain is making people sick. In particular, people lacking resilience are supposedly seeking more psychotherapy due to the human image of some neuroscientists propagated by the media. This has led to disorders of self-perception, anxiety, and the subjective feeling of being completely at the mercy of the brain and its chemistry. In some cases, the idea of *being* the brain has even led to psychoses. If we think this through to the end, the efforts of some modern neuroscientists are indeed intended to make us believe that we live in a world of brains. We ourselves and our fellow human beings would be marionettes of gray cells, which hide in the cave of our skull and pull the strings from there. Each one of our social relationships with another human being or animal would be nothing more than a relationship with another brain.

It is understandable that the idea of a world of brains, void of all minds, can lead to psychoses in unstable people. But worst of all, according to psychotherapists' firsthand experience in private practice, some of their clients suffered a complete breakdown of their sense of purpose when public statements of supposed authorities of brain research were adopted without thinking.

The following condition is most remarkable. Psychiatrists have noticed a frequent occurrence of disorders called "derealization" and "depersonalization." This means that the affected patients no longer perceive themselves and their environment as real. They see houses in the city as two-dimensional scenes—as if they had been painted on cardboard. This is a symptom of derealization. The world suddenly seems like an illusion on a stage. Eventually, many of these patients also see themselves as unreal. This is a symptom of depersonalization. They feel like robots and also perceive other people in a robotic way. When they move, they have the impression that they did not carry out the movement themselves. This can go so far that the feeling of self disappears.

The increase in these disorders can most likely be blamed on—among other things—the worldview of modern neuroscience disseminated by the media. Those representing this view tell us daily that our self as well as the world out there is an illusion that our brain "leads us to believe." Is it really surprising that people actually begin to perceive themselves as unreal and as mere machines under these influences? If we are so sure that we are constructs of the brain, and our consciousness and self do not really exist, why do psychiatrists classify such patients as "sick" and sometimes give them medication? Wouldn't derealization and depersonalization be more of a realization of how the world really is? Obviously, no one ultimately believes that the worldview of this new, materialistic neuroscience is true; otherwise, these psychological disorders would not be regarded as illnesses, but as a kind of "enlightenment."

Interestingly enough, an experience of nature helps against the symptoms of derealization and depersonalization. I personally had a conversation with an affected person who perceived himself as a lifeless robot in a world of props. After weeks of therapy had not helped, he drove to the countryside and spent a lot of time on a river. He was "away" from the impact of society and "away" from his workplace, which was making him sick. He was now surrounded by the healing code of nature, which affected him during the time he spent at the river. His sensory perceptions were activated by feeling the water and stones, listening to the birds, smelling the fragrances, and seeing the plants and animals. This helped him reconnect with the world. Within a few days, he began to "sense" himself as before. He saw himself, the environment, and his fellow human beings as real again. The many sensory impressions in nature accomplished this.[20]

For a long time, we have had to "function" at our workplace like machines in order to serve the higher-ranking matrix—the economy. Nature and processes of life no longer have any value in our society, and economic interests are given precedence—this deceit has been drilled into the minds of humanity. We are uprooted and alienated from our nature. It is no wonder that our "culture of functioning" has defined even humans themselves as born machines.

An important healing code of nature consists of nature helping us to free ourselves from this imprinted materialism and to repair our damaged self-image. As I've written, animals and plants have much more to do with our identity than concrete wastelands, economic interests, and the doctrine of materialism's image of humans. Let us observe wild animals in the forest. They know nothing about the fact that more and more people consider them a mere result of chemical reactions, which scientists might be able to recreate in a test tube one day. Even though we have defined them as something material and banished any life force or spirituality from our thinking, we can recognize anew what life means based on the life and growth processes of animals and plants to which our biophilia responds.

Let's stick to animals a little longer because they are more closely related to us than plants, and we can imagine ourselves in their place more easily. A fox's consciousness is not only in its brain. A fox inhabits its body as a holistic being. Moreover, "the perceptive self does not only live in the body," writes Hans Jürgen Scheurle, "but also in the surrounding space and in things."[21] He is referring to us humans, but it can be applied to animals as well. The life processes of a fox in the forest constantly exceed its body boundaries. The self of a fox lives in the trees that surround it, in the earth into which it has dug its den, in the foliage that it feels under its velvet paws. Its entire identity has sprung from this functional circle with nature and is still embedded in it. The nature around the fox becomes a part of it in a way; nature flows into the formation of its consciousness and becomes the content of its consciousness at the same time. The fox carries an aura, but not in an esoteric sense. The aura of a fox in the forest is the nature in the fox's habitat. The fox becomes aware of objects in the nature surrounding it. They are objects with which it can potentially interact. The tree is not just a tree, but is good for climbing. The brook is there to quench thirst, the hole in the boulder is a good hiding place, and the soft forest floor behind the oak is particularly suitable for digging. The fox's consciousness is constantly expanded by the objects of its habitat.

Besides, the trees are actually there; they are not just an illusion that the brain is leading the fox to believe. The brain does not construct a reality that is not real, but rather *re*constructs a world that the whole body perceives through sensory impressions. When a fox feels the rough surface of a rock under its paws, it is feeling the rock *with its paw* and not in its brain. Neither the rock nor the paw nor the feeling in the paw are constructs of the brain. They are really there. All the sensory stimuli flow into the consciousness of the fox during its foray through the forest. It sees, smells, hears, and feels. Its entire body is active. Perception is always something holistic. The nerves in its paws are active, its eyes perceive colors, its ears hear the sound of the wind, and its nose picks up a scent. Neurons fire in the fox's entire body, and the "abdominal brain" communicates with the brain in its head. The fox's state of consciousness is born as the result of an eco-psychosomatic process.

Considering this complexity, how could anyone still think that consciousness originates only in the brain? The fox's consciousness is also in its paws because there the fox feels the forest floor. Its consciousness is at home in the forest because it sees the red pines where they are, in the forest and not in the brain. Consciousness is connected to the environment, and the fox inhabits its entire body and its environment. It is an eco-psychosomatic unit. The notion that it sees, feels, hears, and smells in the brain is wrong. *We must attribute consciousness to the fox and not its brain.* As I just described, consciousness is the product of a complex eco-psychosomatic process, which is the interaction of a holistic life-form with its environment.

As already described, LUCA's primitive consciousness was the sum of its simple sensory perceptions and its simple environmental interactions. This basic principle has still not changed today. However, the complexity in humans and animals has changed. The consciousness of a human or a fox is a "higher" consciousness than in single-celled organisms because our capability to interact with the environment and maintain social relationships has grown compared to LUCA. Human beings and foxes are, therefore, much more capable of feeling mental and physical pain than an amoeba or volvox sphere. Special ethical principles that we need to adhere to in dealing with other people derive from this ability to suffer. And it goes without saying that our treatment of animals also requires higher ethical principles than those needed for amoeba and volvox. I emphasize this so that I don't give the impression of calling for equality between all beings, from single-celled organisms to human beings.

## Rediscovering the Mind

The great Swiss psychiatrist and psychoanalyst Carl Jung, who died in 1961, wrote that people had become "dirty" through too much civilization and that whenever we came into contact with nature, we would get clean.[22]

Jung, like Einstein, saw a spiritual power in the cosmos and in nature. His entire work as a psychiatrist and psychoanalyst was marked by this conviction. For him, unlike many contemporary neurobiologists, human consciousness could not be reduced to purely material processes. Jung, for example, expressed his love for plants by writing that trees were mysterious and seemed to him to be the embodiment of the hidden meaning of life. For this reason, he felt closest to the deepest meanings and most awe-inspiring effects of life in the woods.[23] Jung sought and found meaning and significance in nature and its life processes.

Viktor Frankl, neurologist, psychiatrist, and founder of logotherapy, complained about "suffering from a life that has become meaningless."[24] The lack of meaning makes us sick, according to Frankl, while a sense of purpose is a prerequisite for mental and physical health. The search for meaning in nature, however, gets difficult or impossible if we reduce nature and even our own consciousness to molecular parts. In this age of the doctrine of materialism, the apparent lack of meaning is certainly one reason for the increase in mental as well as physical disorders.

The German physician and medical professor Arndt Büssing at the Witten/Herdecke University concerns himself with the importance of humans' search for meaning in relation to illness and health. He uses the term "spirituality" in connection with this search for meaning. There is a human primal need that exists in us, detached from religious denominations. Büssing claims that spirituality is the search for sense and meaning and our own origin. This demonstrates that the question of the origin of life is deeply connected with human spirituality. Hence, Büssing's definition of spirituality involves the feeling of being connected with others, with nature, and with the spiritual forces that are active in everything living and in nature.[25] If there is nothing spiritual left in nature because it has been "eliminated," part of the foundation of the human need for spirituality is stripped away.

Spirituality is entering a state of crisis in two ways. Spirituality manifests itself in a person's process of development and consciousness. If we believe in the doctrine of materialism, consciousness, too, becomes a mere illusion. In this view of the world, nature is deprived of the spiritual and humans of consciousness. Both are the basic ingredients for a spiritual sense of purpose.

But let us not forget one thing: despite all attempts of modern sciences to explain life and consciousness on the purely material level, this has not been achieved so far. It remains in countless promises that have not been kept. Lawrence Krauss could not demonstrate that a universe can emerge "out of nothing," nor has a biologist succeeded in producing life out of inanimate matter with mere chemical reactions. Up until the 1980s, researchers broadcast publicly that they would soon be able to explain the origin of life through pure "primordial soup chemistry," but it has been quiet in this field—for decades.

We keep hearing over the years the self-assured predictions of neuroscientists that they will be able to read people's thoughts. Brain research, however, is far from able to explain conclusively how subjective thoughts and feelings are supposed to arise from pure chemistry. It is not enough to know the physics of the color green. We must also know how green light is processed in our brain. When we see the green leaves of a tree, it is connected with the *experience* of the green. "It" does not see green in us, rather, it is *we* who see green. We learn *how it feels* to see a color, to have sensory impressions, and to have feelings. No neurochemist or neurophysicist in the world can explain this "how."

David Chalmers, professor of philosophy and cognitive science at the Australian National University in Canberra runs the Center for Consciousness. Chalmers is one of the world's most influential contemporary philosophers dealing with the human mind. He is known for distinguishing between easy and hard problems of consciousness research. Relatively easy to solve are mental phenomena that are not associated with subjective experience. How do we process sounds on the physical level? How do we learn? How are movement impulses generated? All of these questions can be answered. At the moment, however, the *hard* problems of consciousness research are completely unsolvable. Whenever we experience something subjective—that is, when we learn how something feels—we face a scientific mystery because chemical reactions cannot feel anything. If humans were merely "machines," they would walk, talk, and sleep; they could do a lot of what we do as conscious people, but there would be no one present who experienced *how it feels* to do all this. The essence of the fox that crosses the forest is also present in its body and in the world. The fox feels the quality of being a fox—it knows *how it feels* to be a fox. It knows how it feels to dig a hole in the forest ground or what the bark of a tree feels like under its paws when climbing. Only this particular fox has access to its inner perspective. No one else will ever be able to enter into the fox's stream of consciousness and *be* this fox, not even if we have understood every detail of the fox's brain chemistry and can reproduce it in the laboratory.

Chalmers, however, does not separate matter from mind and assumes that consciousness is a constant of nature. According to him, matter has both physical and nonphysical properties. The latter manifest themselves in human beings and animals as subjective experiences, which the Australian philosopher calls "qualia."[26]

There are material processes that accompany life, but life does not solely consist of them. Those who think they have understood a tree because they cut it up into the smallest of its parts are mistaken. The glue holding everything together is missing.

Ages of civilizations come and go. A science will exist *after* the doctrine of materialism. In earlier centuries, humans described themselves as mechanical beings made of gears and levers. The state of technology has always been reflected in the human image of the respective time. Today, we live in the computer age and see people as computers. Whoever thinks we have reached the apex of our knowledge of humans and nature should take a look in the history books. Another time will come, and many things will change again. We will find new perspectives on consciousness, life, health. Personally, I am firmly convinced that humanity will return to a holistic view of the world, and this will be an eco-psychosomatic one. Nature will show us how to do it. We have learned and can still learn a lot about ourselves from nature. That is why plants and animals, meadows and wetlands, and mountains and forests will always magically attract us. In this context, nature offers us another important aspect of being away: being away from a society in which we have had to function for a long time like machines and that has even now defined us from birth as machines.

Learning from nature that we are more than machines is perhaps the most important healing code of nature in the twenty-first century.

 **THIS CHAPTER IN A NUTSHELL**

Many well-known scientists have spoken out against a materialist view of nature and life processes, including the zoologist Jakob Johann von Uexküll, who describes the functional circle between living beings and their environment. The German physicist Walter Heitler, who died in 1981, warned against "excessive materialism" that is "putting forth evil buds." Albert Einstein was a pantheist–that is, he believed in a spiritual power present throughout the cosmos and nature. Carl Jung also turned against the doctrine of materialism and recognized in nature a spiritual power.

Many things have turned out to be different from what past scientific experts assumed or predicted. The mainstream approach in biology tries to reduce all life processes to mere chemistry. Some would even like to believe they have found the formative life force *eidos*–which Aristotle believed is in life-forms–in the chemical molecule of DNA. Others see the "unmoved mover," which Aristotle saw behind life processes, in a DNA molecule.

The first living beings from four billion years ago are said to have developed from ancient puddles on stones, clay particles, or beads of oil in the water. The efforts of many biologists to find a purely chemistry-based explanation for the initial spark of life have been unsuccessful for several decades. Similar to biologists' efforts, attempts by physicists to convince us that the genesis of the universe "came out of nowhere" have failed.

Until now, the doctrine of materialism culminated in the assertions of contemporary neuroscientists that human consciousness can be traced exclusively to chemical processes in the brain. The same applies to animal consciousness. We ourselves have been declared identical to our brains, and neuroscientists represent the idea that we are in the cranial cavity and are connected to our perceptive organs through nerve pathways. A relationship between two human beings, or an animal and a human, means that neuronal circuits force a brain into associating with another brain, like a robot. Humans are machines, and their minds the software. In many of us, this new human image triggers feelings of unease and psychological problems, since it involves a serious attack on the human self-image and our sense of purpose.

A "consciousness seed" was already planted in the single-cell LUCA. Consciousness has always been a characteristic of the *entire* living being, which is in a functional circle with its environment. It cannot be reduced to one organ. Using the example of the fox, I explained why consciousness is not generated in the brain.

Since the discovery of the "abdominal brain," we know that emotions and moods emanate from there. The old phrase "gut feeling" was right all along. Consciousness arises through the interaction of a holistic living being with the environment and with other living beings. Consciousness transcends our body's boundaries and also feels at home in the world.

Life remains an unsolved mystery and retains its meaning beyond pure material chemistry. This insight might be the most important healing code of nature in the twenty-first century.

7

# THE FUTURE OF ECO-PSYCHOSOMATICS

## A New Foundation for Eco-Psychosomatics

Even the field of conventional psychosomatics had to convince critics in medicine of its validity, although there is compelling evidence for the inseparable unity of mind and body. Eco-psychosomatics will surely encounter even greater resistance because it sees the human organism as extended beyond its physical boundaries, as I have explained in this book. This is the exact opposite of current medical paradigms that pull everything further and further apart. To understand more details about life, health, and disease, it is of course important to analyze the single parts of a system by reducing the system to individual aspects. But, as I've written, we must not forget to switch back to the holistic level again. We shouldn't get stuck on the reductionist view of the world. Eco-psychosomatics means that we really understand the human organism in its greatest context, integrated into the functional circle with nature.

There is no uniform, scientifically established definition of the term "eco-psychosomatics." While working on my book *The Biophilia Effect*, it occurred to me that this word is a good description for nature-human medicine. Then I found out that it was already in use in the early 1990s. The German psychologist Sigrun Preuss wrote a book titled *Ökopsychosomatik* (Eco-psychosomatics) that is about environmental pollution, but not about the natural healing powers of plants, landscapes, and animals.[1] The book describes the psychological and physical problems triggered by pollutants, noise, "electrosmog," toxins, and other environmental impacts of the modern world. For example, it suggests that electromagnetic radiation is a cause of migraines. Preuss, when using the term "eco-psychosomatics," focused on the harmful effects of the modern environment on humans.

So too does Springer Spektrum's online German dictionary of psychology. It also defines eco-psychosomatics solely as a science concerned with environmental influences that damage the mind and body.[2] I didn't even find the term "eco-psychosomatics" in any English dictionary, but since Spektrum is one of the leading international scientific publishers, we can consider this German definition to be the most "widely" accepted (but it is still practically unknown beyond scientists). The problem with this definition is that it completely ignores the health-promoting potential of contact with nature. With this definition, eco-psychosomatics is nothing more than a synonym of "environmental medicine," a field of medicine concerned almost exclusively with the effects of environmental pollutants. Even though the phrase "environmental medicine" itself should also encompass the healing effects of nature, conventionally it deals only with the role of environmental impacts in the genesis of human disease. We should expect that a science called "eco-psychosomatics" or "eco-psychosomatic medicine" will go far beyond this limited definition.

Eco-psychosomatic research must include the healing potential of contact with the natural world, and not only the harmful effects of pollutants, toxins, or electromagnetic fields. Even when we combine the science of environmental pollutants with the benefits of nature in psychotherapy and psychological healing, we still have only two-thirds of eco-psychosomatics covered. As a biologist, I would regret leaving out the missing cellular, organic, and biochemical components. In this book, it has become clear that we are embedded materially and organically in a functional circle with nature that is essential to our health. Eco-psychosomatics is not complete until we add the approach of biomedical research. This way, all scientists, doctors, and therapists can participate in it. Research methods such as blood sampling directly in forests, savannas, grasslands, and so on, as well as laboratory analyses of changes in our immune system, hormone levels, or the heart that occur during contact with plants or animals have only recently been employed in research on nature-human relationships. In this field, Qing Li in particular and other East Asian physicians and biologists have made pioneering contributions to the concept of forest medicine. Through scientific methods that are recognized in evidence-based medicine, eco-psychosomatics becomes a *new* field of science. Reaching far beyond psychotherapeutic and social pedagogical niches, in which "garden therapies" or "nature therapies" have gained ground in the past, nothing should stand in the way of establishing the new field of eco-psychosomatics on a new, interdisciplinary, and evidence-based foundation.

As the foundation for an interdisciplinary community project of eco-psychosomatic science, I propose the following definition in order to get past the previous definition that reduces eco-psychosomatics to a mere science of toxicology:

 **ECO-PSYCHOSOMATICS** is an evidence-based science of the material as well as the nonmaterial impact of plants, animals, and ecosystems on the physical and mental health of human beings, whose organism is viewed as extending beyond the surface of the skin. Like every living being, humans are embedded in an evolutionary functional circle with our natural habitats.

Eco-psychosomatics explores:

1. The medical impact of contact with nature and animals on the organic level (soma) as well as on the psyche

2. The role of environmental toxins, pollutants, electromagnetic forces, noise, and light pollution in the genesis of physical and mental diseases

3. The role of the absence of bioactive substances and sensory stimuli from nature as a cause of disease

Eco-psychosomatics derives its therapeutic procedures from these findings and then evaluates their effectiveness. Biomedical field and laboratory methods are an integral part of eco-psychosomatic research.

With this proposal, I invite scientists, physicians, and therapists to participate in the further dissemination and establishment of eco-psychosomatic medicine, and to help shape it as a future research discipline. Already existing sciences in which the interactions between humans and nature are examined could then be united under the umbrella of eco-psychosomatics. In addition, this approach should be taken into account in veterinary medicine since animals should be regarded as eco-psychosomatic beings, just like humans. The findings from this book can also be applied in modified form to veterinary medical research.

## An Eco-Psychosomatics Clinic

In the first chapter, I told you how much a forest helped me and my son regenerate our mental energy during a long and exhausting hospital stay. Every day, as we walked through the forest and met other patients, I imagined how hospitals could actively integrate the forest into their future concept of treatment plans. For clinicians, psychologists, and nurses, too, a "forest clinic" would be beneficial because they would have a place to relax on their breaks, and the forest would allow them a chance at "being away" from their stressful daily hospital work.

What I haven't told you so far is that in the forest area next to the hospital there was also an animal sanctuary, where wild animals in need were taken care of. For example, Elsa, a wounded squirrel, lived there. Every time we visited her there, Elsa was curious about us and not at all fearful. She climbed up our arms and sat on our shoulders. During these moments, my son and I were able to leave the onerous hospital stay completely behind us. We were "away." The wild animal sanctuary was home to deer, dormice, hedgehogs, and even foxes. We also met ravens, crows, magpies, and other birds.

Emma, a ewe, was not a wild animal, but a farmer had placed her in the care of the sanctuary because he had to close his farm and did not want to slaughter her. Even a raccoon, which the authorities had removed from an illegal and cruel situation, was cared for at this wild animal sanctuary in the middle of the forest.

Whenever possible, the sanctuary staff releases the animals into the wild again. Animals that aren't able to recover enough to survive in the wild can remain permanently at the sanctuary. For these animals, it is not a problem to have interactions with humans. But those that will be released are kept away from human contact.

I found it useful that this "outdoor hospital for animals" was located right next to the hospital for the people. The location was purely a coincidence, but an eco-psychosomatic clinic of the future could focus on healing humans near the healing of animals. As we already know, encounters with animals have a very positive effect on our self-healing powers. Children, in particular, would feel solidarity with a community of animals that are also injured. The animals are sometimes experiencing pain or have had surgery. They sometimes receive medication and are receiving medical care, too. Children could "share" their own suffering with animals.

In chapter 5, I summarized the studies that demonstrate that a human-animal relationship even increases the chances of surviving a heart attack. Animal-assisted therapy is proven to strengthen the immune system and has already been implemented in numerous clinics. Its success is clearly measurable. Therapy involving animal contact cannot be missing in an eco-psychosomatic hospital. It is essential that strict hygienic guidelines are adhered to, of course, and it is not always possible to use dogs, cats, and other animals as co-therapists in every situation. Some hospital departments are closed even to human visitors for sanitary reasons.

In the United States, there are more than six hundred hospitals where animals have been part of the "staff" for years or even decades. There is not a single case where an infection or disease was transmitted to a patient due to this animal contact.[3] The Harlaching Clinic at the General Hospital in Munich, Germany, has demonstrated for ten years that animals can be included in the operation of a big hospital. There, rabbits, guinea pigs, and dogs are an integral part of the medical treatment concept.[4] I have seen numerous clinics all over the world where plants are also integrated into the therapeutic offerings through, for example, garden therapy. Through gardening, doctors invite nature into the clinic, and with it, the healing code of nature. A garden with trees has a positive effect when patients can look out of their hospital room window and see these trees. As we already know, the view of a tree activates our self-healing powers. Gardens bring the healing substances of nature to the hospital. Trees, shrubs, and other garden plants emit terpenes, which strengthen our immune system. What I haven't even mentioned yet is that the health-promoting terpenes in natural ecosystems also flow from the soil, because microorganisms in the soil release them. The closer to nature the soil in a garden is, the quicker a natural flow of substances occurs from the ground.

A nearly natural hospital garden offers patients enough room to encounter plants. Humanity looks back on a 10,000-year-old symbiosis with garden plants. Since humans have established settlements, they have been living closely with these plants.

Tending to and caring for plants at hospitals is of great therapeutic value, which is why some hospitals have developed their own garden therapy programs. The garden activates our nervous system of rest, thanks to the soothing stimuli of nature. Many clinics where physicians offer garden therapy specialize in neurological rehabilitation. Through simple gardening, patients learn new ways to use their bodies after a stroke or accident. Gardening helps the body to rebuild damaged nerve connections. Elevated garden beds make it possible for people in wheelchairs to also work with plants.

Another area of application of garden therapy is in pain treatment. Roger Ulrich, the researcher who investigated the healing effect of viewing a tree, has shown that garden therapy raises the pain threshold. I have spoken with doctors in hospitals who can confirm that, based on patient records, the use of painkillers, antidepressants, and blood pressure medications declines with garden therapy. Garden therapy is also used in conjunction with cancer treatment to achieve a natural pain relief without any side effects. Particularly positive experiences with garden therapy, as well as with animal-assisted therapy, are found in psychiatric and psychotherapeutic clinics. There is also a lot of evidence supporting the use of gardens for elderly people, which is why geriatric clinics and nursing homes often provide garden therapy to their clients. A large-scale study conducted in Australia showed that people over sixty can reduce the risk of dementia by 36 percent by gardening on a daily basis. In this study, the scientists followed nearly three thousand participants for sixteen years.[5]

Japanese doctors even go into the forest with their patients. Forest medicine is not yet in use in European or US hospitals. There are too few forest areas around most hospitals, and excursions with patients to forest areas would be costly, requiring extra staff. In Japan, there is a better chance of financing forest medicine because it is officially recognized by the state health system.

I can't help thinking that it will be a while before public or private health insurance finances eco-psychosomatic treatments in other countries. The health-care systems of most countries haven't even begun to implement the findings of evolutionary medicine and eco-psychosomatic research.

In the summer of 2015, I visited a hospital surrounded by green space in Austria. This space was going to be sold and used for new construction. It would not be available for patients in the future. A chief doctor told me that he considered natural areas around hospitals superfluous. In his opinion, even pulmonary medicine would be fine without forest walks. Forest air had previously been important in curing lung diseases, but drugs are so far advanced in the twenty-first century that we no longer need forest air to make sick lungs healthy again.

His statement clearly demonstrates how unimaginably far some decision makers in the health-care system are from the idea of eco-psychosomatic medicine. To remove nature from the world of *Homo sapiens*—beings of nature—and to replace it with modern pharmaceuticals is the exact opposite of eco-psychosomatics. This position originates from the doctrine of materialism that has taken root in our society. The fact that it also reflects economic interests and the profit margin of the pharmaceutical industry doesn't need more in-depth discussion at this point. Eco-psychosomatics is different.

## Allergies: Proof of the Healing Code of Nature

In the spring of 2015, I took part in a discussion on the Austrian television show *Stöckl*, where I was also joined by the Austrian novelist Vea Kaiser. In the course of the TV show, she expressed doubts about the healing effects of trees, referring to her own pollen allergy: "For me, at least right now in May, the sight of a tree isn't soothing at all. All it makes me do is worry, 'Where are my allergy pills?'"[6]

People repeatedly use allergies to certain plant substances as an argument against the healing code of nature. But does the existence of allergies really undermine the statements of eco-psychosomatics? Actually, the opposite is true! The origin and increase of allergic reactions is, in fact, more convincing proof that the separation from nature is making us sick and that more contact with nature would lead our society back to a healthier path overall. Allergies to natural substances are due to, among other causes, our alienation from nature.

An allergic reaction is an immune response to a harmless antigen from the environment that is erroneously classified as an intruder by our immune system. These antigens from nature are usually pollen. Our body mistakes these microscopically small structures for pathogens. The pollen is called an "allergen" in this case.

The number of allergy cases has doubled in the industrialized world over the past fifteen years.[7] Asthma and allergic skin reactions are on the rise. The increase in these autoimmune diseases can be attributed mainly to environmental influences and are regarded as lifestyle diseases. The four main environmental factors that lead to their occurrence are these:

- Pollution

- Unnatural diet

- More allergens in the environment

- Altered contact with microorganisms during childhood[8]

Our immune system is no longer adequately "trained." It acquires its abilities mostly in early childhood. That is when we set the course of our immune system for a long time. There is no doubt that the lack of contact with bacteria and other microorganisms can lead to future allergies through exaggerated hygiene. The immune system is training its future abilities with microbes. This principle begins at birth. The bacteria of the vaginal mucosa, with which the newborn comes into contact, are usually the baby's first independent contact with microorganisms. These bacteria colonize the intestines after birth and thus become part of the baby's body—and a part of its immune system as well as its "abdominal brain." Statistically speaking, a cesarean birth increases the risk of the baby developing allergies later because contact with the vaginal flora is absent and the intestine is colonized by the wrong bacteria.[9]

Because humans are in a functional circle with their environment, influences from nature help shape our long-term physical and mental health from the first day of existence. Misunderstood hygiene increases the risk of allergies. Soil, often referred to as "dirt," contains bacteria that can be used to test and develop the child's immune system. From studies, we know that children growing up on farms have a much lower risk of suffering from allergies than city children.[10] The more substances, microbes, and other influences from nature with which the immune system is allowed to familiarize itself in the early years, the sooner it will be able to distinguish between "good" and "bad" over the course of life.

Contact with soil organisms is healthy for a number of reasons that go well beyond the subject of allergies. There are indications that the learning ability and intelligence in mammals, so in humans too, benefit from contact with a bacterium biologists call *Mycobacterium vaccae*. This completely harmless, healthy soil bacterium is used in immunotherapy against allergies, asthma, psoriasis, and neurodermatitis. It sounds unbelievable, but several studies suggest that contact with *Mycobacterium vaccae* alleviates depression and anxiety. Observations in mice have also shown that those animals that came into contact with the bacterium found their way through mazes better than the mice without that exposure.[11]

Currently, scientists are busy researching this microorganism that will likely play an important role in medicine in the future, once its effects are better understood. What is certain is that contact with this soil organism, and other microorganisms from the soil too, is important for the formation of a healthy immune system. If we come into regular contact with these organisms during childhood, the risk of developing allergies decreases.

We also know that the occurrence of allergies is additionally facilitated by environmental pollutants. For example, particles from diesel exhaust can amplify the overreaction to allergens. This fact also shows that an ecological rethinking would make our society healthier.[12]

These examples show that being allergic to pollen and other natural substances is no argument against the healing code of nature. This conclusion would be a complete reversal of the facts. The prevalence of allergies shows, in an enormously effective way, that it is the separation from nature that makes us sick, and contact with nature makes us healthier. The eco-psychosomatic system is therefore resoundingly supported through these correlations.


## Eco-Psychosomatics Without the Rose-Colored Glasses

The German literary critic Hellmuth Karasek, who has since passed away, also took part in the same discussion with the writer Vea Kaiser. I had the honor of talking with him about humans and nature. This former professor of theater objected to my idea of a healing nature because nature, according to him, was a place where people fight for survival. "Survival in a jungle after an airplane crash is very brief," he said.[13] He was, of course, quite right. Besides the fact that we have an overall poor chance of surviving a crash, the jungle would be a dangerous place for us modern humans, who are alienated from the wilderness. The search for food alone is likely to be an obstacle for most of us.

The conversation with Karasek offered me the opportunity to clarify a major misunderstanding. The healing code of nature does not mean that the wilderness has *only* a healing impact on us. It would be naive to ignore the dangers of the jungle. Therefore, eco-psychosomatics does not mean abandoning civilization and "living naked in the forest." Eco-psychosomatics is the evidence-based science of the health-promoting influences of nature on humans. And there are plenty of them. It is, therefore, a question of integrating nature into our health care and medicine so that humans—because we are beings of nature—can be in contact with plants, animals, and intact ecosystems once again. This is also associated with a respectful attitude toward nature and animals. Yes, we could starve or be eaten by a crocodile in the jungle, but that does not change the fact that tree terpenes in the forest air, and in future drugs, as well as the impressions and influences of nature, have an immensely positive effect on our health. The protection and conservation of ecosystems and a dignified approach to our relatives, the animals, should be a matter of high importance for us. This does not change the dangers of a jungle.

Sometimes readers ask me why I consider it necessary at all to scientifically prove the healing power of nature. After all, everyone can intuitively feel how good a forest walk is for us. This is only partially true. Particularly among political decision makers of the health-care system, new ideas are not granted a hearing without hard-hitting evidence. Many scientists and doctors continue to consider the healing code of nature "esoteric nonsense." Apparently, not *all* people feel how good nature is for us. If we want to reform our health-care system and fix our relationship with nature, we urgently need scientific evidence. This will make change possible.

This book demonstrates that the healing code of nature is based on a solid scientific foundation. Eco-psychosomatics is the holistic science of humans, animals, and nature. It has nothing to do with rose-colored glasses. Eco-psychosomatics will revolutionize the health-care system.

 **THIS CHAPTER IN A NUTSHELL**

Eco-psychosomatics is the interdisciplinary exploration of nature-human medicine. The term is not yet scientifically established. My research has shown that the word "eco-psychosomatics" was used in the 1990s as a synonym for "environmental medicine"–that is, for the investigation of harmful environmental influences on the body and the psyche. However, environmental medicine is already responsible for this area of study. What makes eco-psychosomatics special is that it views humans as part of the network of life, and the healing code of nature is key. It is not just about environmental pollution, but above all it is about the physical and mental therapeutic effects of nature on humans.

Hospitals in which garden therapy, forest medicine, or animal-assisted therapy are offered could already be described as eco-psychosomatic hospitals. The eco-psychosomatic clinic of the future will specifically reflect the connection between *Homo sapiens* and nature in their therapeutic offerings.

The fact that allergies to pollen and other plants are increasing is not an argument against the healing power of nature. On the contrary, it is proof that eco-psychosomatics does what it promises. Allergies arise mainly through the separation of people from natural influences during early childhood. One example is soil bacteria, which children encounter while playing in nature: children on farms are shown to develop fewer allergies than children living in cities. Allergic reactions mean that more contact with nature is necessary to make our society healthier. It would be a distortion of the facts to think that a pollen allergy contradicts the healing power of nature.

I am not suggesting we look at nature through "rose-colored glasses," either. Of course, spending time in the wilderness can also be dangerous–think wild animals in the jungle. Eco-psychosomatics does not mean we should abandon civilization. It is about the integration of contact with nature into our everyday life and health-care system. Eco-psychosomatics is also associated with an appreciative attitude toward ecosystems and animals.

# THE WHOLE IS GREATER THAN
# THE SUM OF ITS PARTS

## The Complexity of the World

A good way to approach an understanding of "the world" is the concept of a layered construction in accordance with German philosopher Nicolai Hartmann's "levels of reality."[1] You can distinguish between lower and higher layers, or organizational levels—similar to the various layers in a forest ecosystem. These are not hierarchies in the classical sense, but a highly networked system consisting of "holons."[2] Each whole consists of parts and is itself part of a larger whole. The parts can exist without the whole, but not vice versa. A holon of atoms consists of elementary particles, and a holon of molecules consists of atoms. Without atoms, there are no molecules, but without molecules—at very high temperatures, for example—there are still atoms.

Over the course of chemical and later biological evolution, ever higher organizational levels emerged, such as cells, organisms, cognitive abilities, consciousness, and the level of the spiritual. On each higher level, new features appear that cannot be understood from a lower level. If the well-known phrase "all life is chemistry" is supposed to mean that life without chemical processes is not possible, that is correct. But if this saying is supposed to mean that life is nothing but chemistry, it is fundamentally wrong—because what makes life matter lies beyond chemistry.

Suppressing higher organizational levels and their qualities is a kind of reductionism. In empirical science, essential insights have been gained in narrow subfields by ignoring the overall reality. This is a very useful approach at first glance. It becomes dubious and dangerous when you believe you can improve the world with technology derived from these insights without having understood this complex world at all.

## The Failure of Reductionist Approaches

There are numerous cases of the failure of reductionist concepts. In line with the subject of this book, the following example is worth mentioning.

In the 1990s, the Human Genome Project (HGP) was launched with a huge publicity campaign, declaring it would decode the human genome. It promised human DNA sequencing that would lead to significant advances in the understanding of disease and provide the opportunity to produce "tailor-made" drugs. Pharmaceutical companies invested heavily in this potentially profitable research project, to the tune of billions of dollars.

Human DNA sequencing has long since been completed, but the promises associated with it have (fortunately) not been fulfilled. Although we were expecting 100,000 to 140,000 human genes, the HGP yielded only around 20,000 genes, comparable to fruit flies or roundworms. The concept "one gene to one protein" has been debunked. The procedures are much more complex than anticipated. Instead of providing answers to the questions researchers asked, the HGP raised many new questions and showed that the genetic knowledge with which the project was started was largely inadequate. Even Craig Venter, whose company Celera Corporation saw itself at the forefront of biomedical research a few years ago with its "sequencing robots," had to admit, "In retrospect, our assumptions about the way the genome functions were so naive that it is almost embarrassing."[3]

It has become quiet around the Human Genome Project. The current genomic sequencing data are not good for pharmaceutical business but show in all clarity the untenability of the deterministic dogmas that have not yet been discredited. The central dogma of "information runs from DNA over RNA to a protein and not vice versa" no longer applies. What critics had long emphasized—that the current molecular genetic knowledge is far from sufficient to understand biological functions—was only clearly confirmed a few years ago.[4] Namely, the vast majority of DNA in the cell nucleus (98 percent in humans) does not encode any proteins, something that was previously regarded as their only function. That is why this DNA was regarded as worthless and disrespectfully called "junk DNA," which (nonmolecular) biologists rightly felt was a combination of stupidity and arrogance. Today it is assumed that the "epigenetic" control and regulatory functions take place in this area of the genome.

In the field of epigenetics (inheritance of characteristics that are not defined in the DNA sequence), considerable insights have been gained in recent years. The "inheritance of acquired qualities," long regarded as impossible, has been observed in many instances.

Basic concepts on which genetic engineering was based have thus been refuted, such as the assumption that a gene of organism A would act in a strictly definable way and, when transplanted into organism B, act the same way. Despite a scientific basis that cannot be defended, the commercial application of genetic engineering continues to be developed.

Genetic engineering promises are confronted with unforeseeable, negative consequences. The most serious aspect is perhaps that they attempt in a highly reductionist way to define humans based on their DNA instead of understanding and respecting them as holistic beings in a biological and social environment, something we should also demand for our fellow creatures.

## The Important Connection Between Ecology and Medicine

Criticism of molecular biological advances in medicine is not popular. However, bear in mind that increasing human impoverishment and destitution do not require more elaborate, costly medical procedures, which benefit only the privileged, but environmental strategies that in the long run avoid much more disease and human suffering and do justice to the basic human right to defend life and health.

This book by Clemens Arvay shows new and promising possibilities of environmental health in the form of an eco-medicine that is equally beneficial to humans and to all other life-forms. This comprehensive view not only opens up a wide field of preventive medicine but also brings a better understanding of the entire biosphere as a shared destiny and will help to overcome the anthropocentric narrowing of ethics and more clearly recognize the need to preserve a rich ecosphere.

**Dr. Peter Weish**
Associate professor in human ecology at the
University of Natural Resources and Life Sciences, Vienna
*February 2016*

# ACKNOWLEDGMENTS

In the spring of 2015, I met Ulrich Ehrlenspiel from Random House publishing group in Munich, where he was directing the Riemann publishing house. During our first conversation, it became clear that we are both enthusiastic friends of nature and share a great fascination for the healing powers of plants and animals. We agreed that all the ground-breaking discoveries of nature-human medicine needed to be shared with the general public. We also agreed that we wanted to contribute in the propagation of an eco-psychosomatic understanding of life and health and to build a contemporary foundation for the widely unknown science of eco-psychosomatics. Science-based findings show that humans have always been right when they intuitively feel health-promoting effects in the forests, meadows, mountains, and in encounters with animals. Just as the old idiom ("gut feeling") shows that humans knew there was an abdominal brain, which science has now verified, humans of all cultures were aware of the healing effects of nature. Now, research is teaching us how these effects develop and how great their potential is. I would like to thank Mr. Ehrlenspiel for being so enthusiastic about this issue and believing in my book from the first moment of our collaboration and championing the original German-language edition.

I also wish to express thankful recognition for the team of Sounds True Publishing in Boulder, Colorado, for their great work on the English edition of this book. It was a pleasure for me to cooperate with all the awesome people at Sounds True: Jennifer Brown, Sarah Gorecki, Marjorie Woodall, Chloe Prusiewicz, Wendy Gardner, Kira Roark, and Amy Sinopoli.

Furthermore, Victoria Goodrich Graham did an excellent job in translating my German manuscript into English. I am grateful for her competent work.

My thanks to Thomas Haase, rector of the University College for Agrarian and Environmental Pedagogy in Vienna, who contributed a foreword to this book. His university is one of two in Europe where one can get an academic degree in therapeutic intervention with plants and animals (a master's degree in "Green Care").

I thank the ecologist Peter Weish (University of Natural Resources and Life Sciences in Vienna) for reviewing my manuscript and for his valuable remarks. As the culmination of this book, it is an honor to accept his well-written guest commentary—a contribution to the whole, which is more than the sum of its parts. Last, but not least, I thank you, dear reader, for the trust you have given me by reading this book. I hope you have found our journey to the primeval life to be interesting.

I'll keep you up to date at **clemensarvay.com**.

# NOTES

## EPIGRAPH

1.  Hermann Haken, *Erfolgsgeheimnisse der Natur. Synergetik: Die Lehre vom Zusammenwirken* [The science of structure: Synergetics] (Munich: Deutsche Verlags-Anstalt, 1994), 27. Hermann Haken, born in 1927, is professor emeritus in theoretical physics at the University of Stuttgart.

## CHAPTER 1: THE MYSTERY OF TREES

1.  Roger Ulrich, "View Through a Window May Influence Recovery from Surgery," *Science* 224, no. 4647 (April 27, 1984): 420–21.

2.  Qing Li, Maiko Kobayashi, and Tomoyuki Kawada, "Relationships Between Percentage of Forest Coverage and Standardized Mortality Ratios (SMR) of Cancers in All Prefectures in Japan," *Open Public Health Journal* 1 (2008): 1–7, doi: 10.2174/1874944500801010001.

3.  Omid Kardan et al., "Neighborhood Greenspace and Health in a Large Urban Center," *Scientific Reports* 5, no. 11610 (2015), doi:10.1038/srep11610.

4.  Geoffrey Vendeville, "Living on Tree-Lined Streets Has Health Benefits, Study Finds," *Toronto Star*, July 13, 2015, thestar.com/news/gta/2015/07/13/living-on-tree-lined-streets-has-health-benefits-study-finds.html.

5.  Vendeville, "Living on Tree-Lined Streets."

6. You can find more information about *shinrin-yoku*, "forest bathing," in Clemens G. Arvay, *The Biophilia Effect: A Scientific and Spiritual Exploration of the Healing Bond Between Humans and Nature* (Boulder, CO: Sounds True, 2018), 11, 77–78, 154.

7. Qing Li and Tomoyuki Kawada, "Effect of Forest Environment on Human Immune Function," in *Forest Medicine*, ed. Qing Li (New York: NOVA Biomedical, 2013), 69–89.

8. Volker Tschuschke, *Psychoonkologie: Psychologische Aspekte der Entstehung und Bewältigung von Krebs* [Psychooncology: Psychological aspects of the development and management of cancer] (Stuttgart: Schattauer Verlag, 2011), 81, 116.

9. Kenneth M. Murphy, Paul Travers, and Mark Walport, *Janeway Immunologie* [Janeway's immunobiology] (Berlin: Springer Spektrum, 2014), 123.

10. The three most important anticarcinogenic proteins, which multiply in forest air, are perforin, granulysin, and granzyme.

11. Li and Kawada, "Effect of Forest Environment," 74–86.

12. Tatsuro Ohira and Naoyuki Matsui, "Phytoncides in Forest Atmosphere," in *Forest Medicine*, 27–36.

13. Qing Li et al., "Effect of Phytoncides from Forest Environments on Immune Function," in *Forest Medicine*, 162–65.

14. Li et al., "Effect of Phytoncides," 160.

15. The strongest ones turned out to be isoprene, alpha-pinene, beta-pinene, d-limonene and other limonene, as well as 1, 8 cineole among others.

16. Roslin Thoppil and Anupam Bishayee, "Terpenoids as Potential Chemopreventative and Therapeutic Agents in Liver Cancer," *World Journal of Hepatology* 3, no. 9 (2011): 228–49, doi:10.4254/wjh.v3.i9.228.

17. S. L. Da Silva, P. M. Figueiredo, and T. Yano, "Chemotherapeutic Potential of the Volatile Oils from *Zanthoxylum rhoifolium* Lam Leaves," *European Journal of Pharmacology* 576, nos. 1–3 (2007): 180–88, doi:10.1016/j.ejphar.2007.07.065.

18. Forest medicine and laboratory studies by international cancer researchers discovered d-limonene to be the most effective tested substance against tumors.

19. Dietrich Wabner and Christine Beier, eds., *Aromatherapie: Grundlagen, Wirkprinzipien, Praxis* [Aromatherapy: Basics, active principles, practice] (Munich: Urban & Fischer/Elsevier, 2011), 496; and Pamela L. Crowell, "Monoterpenes in Breast Cancer Chemoprevention," *Journal of Breast*

*Cancer Research and Treatment* 46, nos. 2–3 (November/December 1997): 191–97.

20. Wabner and Beier, *Aromatherapie*, 496; and J. J. Mills et al., "Induction of Apoptosis in Liver Tumors by the Monoterpene Perillyl Alcohol," *Journal of Cancer Research* 55, no. 5 (March 1995): 979–83.

21. Wabner and Beier, *Aromatherapie*, 64.

22. Li and Kawada, "Effect of Forest Environment," 70.

23. Joel E. Dimsdale, foreword to *Psychoneuroimmunologie und Psychotherapie*, ed. Christian Schubert (Stuttgart: Schattauer Verlag, 2011), v.

## CHAPTER 2: EVOLUTION AND MEDICINE

1. Raine Sihvonen et al., "Arthroscopic Partial Meniscectomy Versus Sham Surgery for a Degenerative Meniscal Tear," *New England Journal of Medicine* 369 (2013): 2515–24, doi: 10.1056/NEJMoa1305189.

2. M. Englund and L. S. Lohmander, "Risk Factors for Symptomatic Knee Osteoarthritis Fifteen to Twenty-Two Years After Meniscectomy," *Arthritis & Rheumatology* 50, no. 4 (September 2004): 2811–19, doi:10.1002/art.20489; and E. M. Roos et al., "Long-Term Outcome of Meniscectomy: Symptoms, Function, and Performance Tests in Patients With or Without Radiographic Osteoarthritis Compared to Matched Controls," *Osteoarthritis and Cartilage* 9, no. 4 (May 2001): 316–24, doi:10.1053/joca.2000.0391.

3. Theodosius Dobzhansky, "Nothing in Biology Makes Sense Except in the Light of Evolution," *American Biology Teacher* 35, no. 3 (March 1973): 125–29, doi: JSTOR 4444260.

4. "HI-Virus Entkommt Durch Schnelle Mutation," *Spiegel Online*, February 26, 2009, spiegel.de/wissenschaft/mensch/highspeed-evolution-hi-virus-entkommt-durch-schnelle-mutation-a-610061.html.

5. Werner Buselmaier, *Evolutionäre Medizin: Eine Einführung für Mediziner und Biologen* [Evolutionary medicine: An introduction for physicians and biologists] (Wiesbaden: Springer, 2015), 12.

6. Qing Li and Tomoyuki Kawada, "Effect of Forest Environment on Human Immune Function," in *Forest Medicine*, ed. Qing Li (New York: NOVA Biomedical, 2013), 111–35.

7. Steven Stearns and Jacob Koella, *Evolution in Health and Disease* (New York: Oxford University Press, 2008), 279.

8. Colloquium discussion as presented in Carola Otterstedt and Michael Rosenberger, eds., *Gefährten—Konkurrenten—Verwandte: Die*

*Mensch-Tier-Beziehung im wissenschaftlichen Diskurs* [Companions, competitors, relatives: The human-animal relationship in scientific discourse] (Göttingen: Vandenhoeck & Ruprecht, 2009), 153–57.

9.  Edward Wilson, *Die soziale Eroberung der Erde: Eine biologische Geschichte des Menschen* [The social conquest of earth] (Munich: C. H. Beck, 2013), 221.

10. Petra Paumkirchner, "Warum Kratzgeräusche in den Ohren schmerzen" [Why scratching sounds hurt ears], *Die Presse*, December 10, 2011, diepresse.com/home/science/715705/print.do.

## CHAPTER 3: HUMANS IN THE NETWORK OF LIFE

1.  Qing Li et al., "Physical Factors in the Forest Environment," in *Forest Medicine*, ed. Qing Li (New York: NOVA Biomedical, 2013), 162–65.

2.  Marcus Pembrey et al., "Sex-Specific, Male-Line Transgenerational Responses in Humans," *European Journal of Human Genetics* 14 (2006): 159–66, doi:10.1038/sj.ejhg.5201538.

3.  Marcus Pembrey, "Time to Take Epigenetic Inheritance Seriously," *European Journal of Human Genetics* 10 (2002): 669–71, doi:10.1038/sj.ejhg.5200901.

4.  Jonathan Seckl, interviewed in "The Ghost in Your Genes," on BBC's *Horizon*, aired November 3, 2005, available at dailymotion.com/video/x2mwr5i.

5.  Marcus Pembrey, interviewed in "The Ghost in Your Genes," on BBC's *Horizon*, aired November 3, 2005, available at dailymotion.com/video/x2mwr5i.

6.  Alva Noë, *Du bist nicht dein Gehirn: Eine radikale Philosophie des Bewusstseins* [Out of our heads: Why you are not your brain, and other lessons from the biology of consciousness] (Munich: Piper, 2010), 89.

7.  Wikipedia, s.v. "Gaia-Hypothese" [Gaia hypothesis], last modified March 4, 2017, de.wikipedia.org/wiki/Gaia-Hypothese.

8.  Matthew Botvinick and Jonathan Cohen, "Rubber Hands 'Feel' Touch That Eyes See," *Nature* 391 (February 1998): 756, doi: 10.1038/35784.

9.  Noë, *Du bist nicht dein Gehirn* [Out of our heads], 93.

10. Stuart Jeffries, "Neil Harbisson: The World's First Cyborg Artist," *Guardian*, May 6, 2014, theguardian.com/artanddesign/2014/may/06/neil-harbisson-worlds-first-cyborg-artist.

11. Jim Hazelwood, "Mit der Haut sehen: Eine Orientierungshilfe für Blinde " [See with the skin], *Die Zeit*, May 12, 1967.

12. Erich Fromm, *Die Seele des Menschen: Ihre Fähigkeit zum Guten und zum Bösen* [The heart of man: Its genius for good and evil] (Stuttgart: Deutsche Verlagsanstalt, 1979), 42.

13. Fromm, *Die Seele des Menschen*, 43.

14. Edward Wilson, *Biophilia: The Human Bond with Other Species* (Cambridge, MA: Harvard University Press, 1986), 1.

15. Gordon Orians and Judith Heerwagen, "Humans, Habitats and Aesthetics," in *The Biophilia Hypothesis*, ed. Stephen R. Kellert and Edward O. Wilson (Washington, DC: Island Press/Shearwater, 1993), 157–63.

16. John Falk and John D. Balling, "Evolutionary Influence on Human Landscape Preference," *Environment and Behavior* 42, no. 4 (2010): 479–93, doi:10.1177/0013916509341244.

17. Michael Jackson, interview by Martin Bashir, broadcast on ITV2, February 3, 2003, transcript at mjadvocate.blogspot.com/2010/11/living-with-michael-jackson-part-2-of-9.html.

## CHAPTER 4: ECO-PSYCHOSOMATICS

1. Qing Li et al., "Effects of Forest Environments on Cardiovascular and Metabolic Parameters," in *Forest Medicine*, ed. Qing Li (New York: NOVA Biomedical, 2013), 124.

2. Hiroaki Kawano et al., "Dehydroepiandrosterone Supplementation Improves Endothelial Function and Insulin Sensitivity in Men," *Journal of Clinical Endocrinology and Metabolism* 88, no. 7 (July 2003): 3190–95, doi:10.1210/jc.2002-021603.

3. For additional studies, see ncbi.nlm.nih.gov/pmc/?term=DHEA.

4. Nicole Maninger et al., "Neurobiological and Neuropsychiatric Effects of Dehydroepiandrosterone (DHEA) and DHEA sulfate (DHEAS)," *Frontiers in Neuroendocrinology* 30, no. 1 (January 2009): 65–91, doi: 10.1016/j.yfrne.2008.11.002.

5. This degenerative process is lipid peroxidation—that is, an oxidative degradation of lipids.

6. Nikolaos Samaras et al., "A Review of Age-Related Dehydroepiandrosterone Decline and Its Association with Well-Known Geriatric Syndromes: Is Treatment Beneficial?" *Rejuvenation Research* 16, no. 4 (August 2013): 285–94, doi: 10.1089/rej.2013.1425.

7. Wikipedia, s.v. "Yin und Yang" [Yin and yang], last modified January 12, 2018, 18:43, de.wikipedia.org/wiki/Yin_und_Yang.

8. Summarized from Mark Bear, Barry Connors, and Michael Paradiso, *Neurowissenschaften: Ein grundlegendes Lehrbuch für Biologie, Medizin und Psychologie* [Neuroscience: Exploring the brain] (Berlin/Heidelberg: Springer Spektrum, 2012), 547–52.

9. Kurt Kotrschal, "Die evolutionäre Theorie der Mensch-Tier-Beziehung" [Evolutionary theory of human-animal relationship], in *Gefährten—Konkurrenten—Verwandte: Die Mensch-Tier-Beziehung im wissenschaftlichen Diskurs*, ed. Carola Otterstedt and Michael Rosenberger (Gottingen: Vandenhoeck & Ruprecht, 2009), 63.

10. Bum-Jin Park et al., "Effect of Forest Environment on Physiological Relaxation Using the Results of Field Tests at 35 Sites Throughout Japan," in *Forest Medicine*, 57–67.

11. Li et al., "Effects of Forest Environments on Cardiovascular and Metabolic Parameters," 117–35.

12. Renate Cervinka, professor of psychology at the Medical University of Vienna, speaking on the Austrian podcast "Ö1-Radiokolleg" on Österreichischer Rundfunk, November 30, 2015.

13. Rodney Matsuoka, "High School Landscapes and Student Performance" (PhD diss., University of Michigan, Ann Arbor, 2008), 78–92, deepblue. lib.umich.edu/handle/2027.42/61641.

14. Richard Louv, *Das Prinzip Natur: Grünes Leben im digitalen Zeitalter* [The nature principle: Reconnecting with life in a virtual age], (Weinheim: Beltz, 2012), 50.

15. Summarized from Clare Cooper Marcus and Naomi A. Sachs, *Therapeutic Landscapes: An Evidence-Based Approach to Designing Healing Gardens and Restorative Outdoor Spaces* (Hoboken, NJ: John Wiley & Sons, 2013), 14–46.

16. Bear, Connors, and Paradiso, *Neurowissenschaften*, 548.

17. Henry David Thoreau, "Walking," in *Civil Disobedience and Other Essays* (Mineola, NY: Dover, 1993), 63.

18. William Borrie, Angela Meyer, and Ian Foster, "Wilderness Experiences as Sanctuary and Refuge from Society," in US Department of Agriculture, *Wilderness Visitor Experiences: Progress in Research and Management*, comp. David Cole, Proc. RMRS-P-66 (Fort Collins, CO: US Dept. of Agriculture, Forest Service, Rocky Mountain Research Station, 2011), 70–76.

19. US Department of Agriculture, *Wilderness Visitor Experiences*.

20. You can find more information about nature as a psychotherapist in Clemens G. Arvay, *The Biophilia Effect: A Scientific and Spiritual*

*Exploration of the Healing Bond Between Humans and Nature* (Boulder, CO: Sounds True, 2018), 69–73.

21. Jean-Jacques Rousseau, *Die Bekenntnisse* [The Confessions] (Munich: Deutscher Taschenbuch Verlag, 2012), 162.

## CHAPTER 5: THE HEALING BOND BETWEEN HUMANS AND ANIMALS

1. Alison Abbot, "Neuroscience: The Rat Pack," *Nature* 465 (2010): 282–83, doi: 10.1038/465282a.

2. "Ratte verarbeitet Information so gut wie der Mensch" [Rat processes information as well as humans], *Welt*, March 14, 2012, welt.de/wissenschaft/article13921307/Ratte-verarbeitet-Information-so-gut-wie-der-Mensch.html.

3. Statement on Anne Churchland's faculty web page, Cold Spring Harbor Laboratory, accessed November 20, 2015, cshl.edu/research/faculty-staff/anne-churchland/.

4. Kenneth S. Saladin, *Anatomy & Physiology: The Unity of Form and Function* (New York: McGraw-Hill Education, 2014), 7.

5. Henri Julius et al., *Bindung zu Tieren: Psychologische und neurobiologische Grundlagen tiergestützter Interventionen* [Attachment to pets: An integrative view of human-animal relationships with implications for therapeutic practice] (Göttingen: Hogrefe, 2014), 31.

6. Julius et al., *Bindung zu Tieren*, 32.

7. Wolf Herre and Manfred Röhrs, *Haustiere: zoologisch gesehen* [Pets: Seen zoologically] (Berlin/Heidelberg: Springer Spektrum, 2013), 136.

8. Kurt Kotrschal, C. Schlögl, and T. Bugnyar, "Dumme Vögel?—Lektionen von Rabenvögeln und Gänsen" [Stupid birds? Lessons of ravens and geese], *Biologie in unserer Zeit* 6 (2007): 366–74; Irene Pepperberg, *The Alex Studies: Cognitive and Communicative Abilities of Grey Parrots* (Cambridge, MA: Harvard University Press, 1999); and Nathan Emery and Nicola Clayton, "The Mentality of Crows: Convergent Evolution of Intelligence in Corvids and Apes," *Science* 306, no. 5703 (December 2004): 1903–07, doi: 10.1126/science.1098410.

9. Carl Charnetski, Sandra Riggers, and Francis Brennan, "Effect of Petting a Dog on Immune System Function," *Psychological Reports* 95, no. 3 pt. 2 (2004): 1087–91, doi: 10.2466/pr0.95.3f.1087-1091.

10. Julius et al., *Bindung zu Tieren*, 63.

11. Janelle Nimer and Brad Lundahl, "Animal-Assisted Therapy: A Meta-Analysis," *Anthrozoös* 20, no. 3 (2007): 225–38, doi: 10.2752/089279307X224773.

12. Bruce Headey, "Health Benefits and Health Cost Savings Due to Pets: Preliminary Estimates from an Australian National Survey," *Social Indicators Research* 47, no. 2 (1999): 233–43, doi: 10.1023/A:1006892908532.

13. Erika Friedmann, Sue Thomas, and T. J. Eddy, "Companion Animals and Human Health: Physical and Cardiovascular Influences," in *Companion Animals and Us: Exploring the Relationships Between People and Pets*, ed. Anthony L. Podberscek, Elizabeth Paul, and James A. Serpell (Cambridge, MA: Cambridge University Press, 2000), 125–42; Erika Friedmann and Sue Thomas, "Pet Ownership, Social Support, and One-Year Survival after Acute Myocardial Infarction in the Cardiac Arrhythmia Suppression Trial (CAST)," in *Companion Animals in Human Health*, ed. Cindy Wilson and Dennis C. Turner (Thousand Oaks, CA: Sage, 1998), 187–201; and Erika Friedmann et al., "A Friendly Dog as Potential Moderator of Cardiovascular Response to Speech in Older Hypertensives," *Anthrozoös* 20, no. 1 (2007): 51–63, doi:10.2752/089279307780216605.

14. Andreas Zieger, "Erfahrungen mit Tieren in der Betreuung von schwersthirngeschädigten Menschen im Koma und Wachkoma und ihren Angehörigen" [Experiences with animals in the care of severely brain damaged people in coma and vegetative state and their relatives], in *Menschen brauchen Tiere: Grundlagen und Praxis der tiergestützten Pädagogik und Therapie*, ed. Erhard Olbrich and Carola Otterstedt (Stuttgart: Kosmos, 2003), 214–27.

15. Andreas Zieger's Projekt Tierbesuch [Project animal] web page, accessed November 22, 2015, a-zieger.de/Dateien/Vortraege/FolienVortragEschwege_R2006.pdf.

16. Stefanie Böttger, "Die Mensch-Tier-Beziehung aus neuropsychologischer Perspektive am Beispiel der tiergestützten Therapie" [The human-animal relationship from a psychological perspective using animal-assisted therapy], in *Gefährten—Konkurrenten—Verwandte*, ed. Carola Otterstedt and Michael Rosenberger (Gottingen: Vandenhoeck & Ruprecht, 2009), 90.

17. Yoram Barak et al., "Animal-Assisted Therapy for Elderly Schizophrenic Patients: A One-Year Controlled Trial," *American Journal of Geriatric Psychiatry* 9, no.4 (2001): 439–42, doi: 10.1097/00019442-200111000-00013.

18. Carolyn Marr et al., "Animal-Assisted Therapy in Psychiatric Rehabilitation," *Anthrozoös* 13, no. 1 (2000): 43–47, doi: 10.2752/089279300786999950.

19. Bente Berget, Øivind Ekeberg, and Bjarne Braastad, "Animal-Assisted Therapy with Farm Animals for Persons with Psychiatric Disorders: Effects on Self-Efficacy, Coping Ability and Quality of Life, a Randomized Controlled Trial," *Clinical Practice and Epidemiology in Mental Health* 4,

no. 9 (2008), doi:10.1186/1745- 0179-4-9; and Jan Hassink and Majken van Dijk, *Farming for Health: Green-Care Farming Across Europe and the United States of America* (Dordrecht: Springer, 2008).

20. Bente Berget et al., "Animal-Assisted Therapy with Farm Animals for Persons with Psychiatric Disorders: Effects on Anxiety and Depression, a Randomized Controlled Trial," *Occupational Therapy in Mental Health* 27, no. 1 (2011): 50–64.

21. Aubrey Fine, ed., *Handbook of Animal-Assisted Therapy: Foundations and Guidelines for Animal-Assisted Interventions* (London: Elsevier, 2015), 118–21.

22. Fine, *Handbook of Animal-Assisted Therapy*, 195–205.

23. L. Havener et al., "The Effects of a Companion Animal on Distress in Children Undergoing Dental Procedures," *Issues in Comprehensive Pediatric Nursing* 24, no. 2 (April–June 2001): 137–52.

24. Lynette Hart and Mariko Yamamoto, "Recruiting Psychosocial Health Effects of Animals for Families and Communities: Transition to Practice," in *Handbook of Animal-Assisted Therapy*, 56, 58.

25. Erhard Olbrich, "Bausteine einer Theorie der Mensch-Tier-Beziehung" [Elements of a theory of the human-animal relationship], in *Gefährten—Konkurrenten—Verwandte*, 116–17.

26. For the full interview with Andreas Danzer, see Clemens G. Arvay, *The Biophilia Effect: A Scientific and Spiritual Exploration of the Healing Bond Between Humans and Nature* (Boulder, CO: Sounds True, 2018), 1–3.

27. Elisa Sobo, Brenda Eng, and Nadine Kassity-Krich, "Canine Visitation (Pet) Therapy: Pilot Data on Decreases in Child Pain Perception," *Journal of Holistic Nursing* 24, no. 1 (March 2006): 51–57.

28. Julius et al., *Bindung zu Tieren*, 90.

## CHAPTER 6: THE BIG SECRET OF LIFE

1. Joachim Bauer, *Das kooperative Gen: Evolution als kreativer Prozess* [The cooperative gene: Evolution as a creative process] (Munich: Heyne, 2010).

2. James A. Shapiro, *Evolution: A View from the 21st Century* (Upper Saddle River, NJ: Prentice Hall, 2011).

3. Heinz Penzlin, *Das Phänomen Leben: Grundfragen der theoretischen Biologie* [The phenomenon of life: Basic questions of theoretical biology] (Heidelberg: Springer Spektrum, 2013), 8.

4. Henri Julius et al., *Bindung zu Tieren: Psychologische und neurobiologische Grundlagen tiergestützter Interventionen* (Göttingen: Hogrefe, 2014).

5. Penzlin, *Das Phänomen Leben*, 9.

6. Veiko Krauss, *Gene, Zufall, Selektion: Populäre Vorstellungen zur Evolution und der Stand des Wissens* [Genes, chance selection: Popular ideas on evolution and the state of knowledge] (Heidelberg: Springer Spektrum, 2014), 20.

7. David Albert, "On the Origin of Everything," *New York Times*, March 23, 2012, nytimes.com/2012/03/25/books/review/a-universe-from-nothing-by-lawrence-m-krauss.html?pagewanted=all&_r=0.

8. According to this theory, an RNA molecule with a membrane was the first living being. RNA (ribonucleic acid) is still involved in reading and writing the genetic code of cells today.

9. Thomas Fuchs, *Das Gehirn, ein Beziehungsorgan: Eine phänomenologisch-ökologische Konzeption* [The brain: A mediating organ] (Stuttgart: Kohlhammer, 2013), 46.

10. Jan Völker, *Ästhetik der Lebendigkeit: Kants dritte Kritik* [Aesthetics of aliveness: Kant's third critique] (Paderborn: Wilhelm Fink Verlag, 2011), 131.

11. All Einstein quotations come from Albert Einstein, *Mein Weltbild* [The world as I see it] (Frankfurt am Main: Ullstein, 1981).

12. Ernst Pöppel, "Wir sind gemeint: Neurowissenschaftliche und evolutionstheoretische Aspekte des Bewusstseins auf der Grundlage eines pragmatischen Monismus" [We are meant: Neuroscience and evolutionary aspects of consciousness on the basis of a pragmatic monism], in *Spiritualität transdisziplinär. Wissenschaftliche Grundlagen im Zusammenhang mit Gesundheit und Krankheit* [Transdisciplinary spirituality: Scientific foundations in health and disease], ed. Arndt Büssing and Niko Kohls (Berlin/Heidelberg: Springer, 2011), 3–21.

13. Michael Schmidt-Salomon, *Jenseits von Gut und Böse: Warum wir ohne Moral die besseren Menschen sind* [Beyond good and evil: Why we are better people without morality] (Munich: Piper, 2012), 240–46.

14. Diane Ackerman, *A Natural History of the Senses* (New York: Vintage, 1991), xix.

15. Hans Jürgen Scheurle, *Das Gehirn ist nicht einsam: Resonanzen zwischen Gehirn, Leib und Umwelt* [The brain is not lonely: Resonances between brain, body, and environment] (Stuttgart: Kohlhammer, 2013), 22.

16. Pöppel, "Wir sind gemeint," 4.

17. Walter Heitler, *Naturwissenschaft Ist Geisteswissenschaft* [Science is spiritual science] (Zürich: Die Waage, 1972), 14.

18. Scheurle, *Das Gehirn ist nicht einsam*, 24.

19. Gerhard Roth, *Fühlen, Denken, Handeln: Wie das Gehirn unser Verhalten steuert* [Feeling, thinking, acting: How the brain controls our behaviour] (Frankfurt: Suhrkamp, 2003).

20. You can find the full interview with patient "John" in Clemens G. Arvay, *The Biophilia Effect: A Scientific and Spiritual Exploration of the Healing Bond Between Humans and Nature* (Boulder, CO: Sounds True, 2018), 120–24.

21. Scheurle, *Das Gehirn ist nicht einsam*, 35.

22. C. G. Jung, *The Earth Has a Soul: C. G. Jung on Nature, Technology and Modern Life*, ed. Meredith Sabini (Berkeley, CA: North Atlantic Books, 2008), 22.

23. Jung, *The Earth Has a Soul*, 29.

24. Viktor Frankl, "Das Leiden am sinnlos gewordenen Leben" [The suffering of meaningless life], lecture, Vienna, March 25, 1976, tape on core (AEG), MP3 copy, 0:2:37, mediathek.at/atom/14836C5C-00F-001A1-00000948-148278B6.

25. Ralph Marc Steinmann, "Zur Begriffsbestimmung von Spiritualität," in Büssing and Kohls, eds., *Spiritualität transdisziplinär*, 42.

26. David Chalmers, *The Character of Consciousness* (New York: Oxford University Press, 2010), 342–43.

## CHAPTER 7: THE FUTURE OF ECO-PSYCHOSOMATICS

1. Sigrun Preuss, *Ökopsychosomatik: Umweltbelastungen und psychovegetative Beschwerden* [Ecopsychosomatic: Environmental pollution and psycho-vegetative complaints] (Heidelberg: Roland Asanger Verlag, 1995).

2. Spekrum.de Lexicon der Psychologie, s.v. "Ökopsychosomatik" [Eco-psychosomatics], accessed January 7, 2016, spektrum.de/lexikon/psychologie/oekopsychosomatik/10840.

3. Anke Prothmann, "Tiergestützte Interventionen in der Humanmedizin" [Animal assisted intervention in human medicine] in *Gefährten—Konkurrenten—Verwandte*, ed. Carola Otterstedt and Michael Rosenberger (Gottingen: Vandenhoeck & Ruprecht, 2009),193.

4. Stefanie Böttger, "Die Mensch-Tier-Beziehung aus neuropsychologischer Perspektive am Beispiel der tiergestützten Therapie," in *Gefährten—Konkurrenten—Verwandte*, 84–89.

5. Richard Louv, *Das Prinzip Natur: Grünes Leben im digitalen Zeitalter* [The nature principle: Reconnecting with life in a virtual age] (Basel: Beltz, 2012), 102.

6. Vea Kaiser, interviewed by Barbara Stöckl on the television talk show *Stöckl*, aired May 28, 2015 on ORF (Austrian national public service broadcaster).

7. Kenneth M. Murphy, Paul Travers, and Mark Walport, *Janeway Immunologie* (Berlin: Springer Spektrum, 2014), 700.

8. Murphy, Travers, and Walport, *Janeway Immunologie*, 708.

9. H. Renz-Polster et al., "Caesarean Section Delivery and the Risk of Allergic Disorders in Childhood," *Clinical and Experimental Allergy* 35, no. 11 (2005): 1466–72, doi: 10.1111/j.1365-2222.2005.02356.x.

10. Lajos Schöne, "Was bei Kindern wirklich Allergien auslöst" [What really triggers allergies in children], *Welt*, July 15, 2015, welt.de/gesundheit/article144000974/Was-bei-Kindern-wirklich-Al-lergien-ausloest.html.

11. Dorothy Matthews and Susan Jenks, "Ingestion of Mycobacterium Vaccae Decreases Anxiety-Related Behavior and Improves Learning in Mice," *Journal of Behavioural Processes* 96 (June 2013): 27–35, doi:10.1016/j.beproc.2013.02.007.

12. Murphy, Travers, and Walport, *Janeway Immunologie*, 708–9.

13. Hellmuth Karasek, interviewed by Barbara Stöckl on the television talk show *Stöckl*, aired May 28, 2015 on ORF (Austrian national public service broadcaster).

## AFTERWORD: THE WHOLE IS GREATER THAN THE SUM OF ITS PARTS

1. Nicolai Hartmann, *Der Aufbau der realen Welt* [Building the real world] (Berlin: Walter de Gruyter, 1964).

2. Hungarian-British author and journalist Arthur Koestler coined the term "holon" (Greek for "whole") in his 1967 book, *The Ghost in the Machine*, for something that is both a whole and a part.

3. Craig Venter, quoted in B. Kegel, *Epigenetik: Wie Erfahrungen vererbt werden* [Epigenetics: How experiences are inherited] (Köln: DuMont, 2009), 56. These "embarrassing" assumptions are still part of the "scientific" foundation for genetic practice today.

4. See example from W. W. Gibbs, "The Unseen Genome: Gems Among the Junk," *Scientific American*, November 2003, 14.

# INDEX

# ABOUT THE AUTHOR

Born in 1980, **Clemens G. Arvay** is an Austrian biologist. He studied landscape ecology (BSc) at Graz University and applied plant sciences (MSc) at the University of Natural Resources and Life Sciences in Vienna. Arvay examines the relationship between humans and nature, focusing on the health-promoting effects of contact with plants, animals, and landscapes. The author also addresses a second range of topics that includes ecologically produced food along with the economics of large food conglomerates. Clemens G. Arvay has written numerous books, including his German bestseller *The Biophilia Effect*. Learn more at clemensarvay.com

# ABOUT SOUNDS TRUE

Sounds True is a multimedia publisher whose mission is to inspire and support personal transformation and spiritual awakening. Founded in 1985 and located in Boulder, Colorado, we work with many of the leading spiritual teachers, thinkers, healers, and visionary artists of our time. We strive with every title to preserve the essential "living wisdom" of the author or artist. It is our goal to create products that not only provide information to a reader or listener, but that also embody the quality of a wisdom transmission.

For those seeking genuine transformation, Sounds True is your trusted partner. At SoundsTrue.com you will find a wealth of free resources to support your journey, including exclusive weekly audio interviews, free downloads, interactive learning tools, and other special savings on all our titles.

To learn more, please visit SoundsTrue.com/freegifts or call us toll-free at 800.333.9185.